Emerging Materials for Energy Applications

Emerging Materials for Energy Applications

Guest Editors

Fernando B. Naranjo
Susana Fernández

MDPI Basel • Beijing • Wuhan • Barcelona • Belgrade • Novi Sad • Cluj • Manchester

Guest Editors

Fernando B. Naranjo
Electronics Department
University of Alcalá
Alcalá de Henares
Spain

Susana Fernández
Centro de Investigaciones
Energéticas
Medioambientales y
Tecnológicas
Madrid
Spain

Editorial Office

MDPI AG
Grosspeteranlage 5
4052 Basel, Switzerland

This is a reprint of the Special Issue, published open access by the journal *Materials* (ISSN 1996-1944), freely accessible at: https://www.mdpi.com/journal/materials/special_issues/emerging_mat_energy.

For citation purposes, cite each article independently as indicated on the article page online and as indicated below:

Lastname, A.A.; Lastname, B.B. Article Title. *Journal Name* **Year**, *Volume Number*, Page Range.

ISBN 978-3-7258-7216-9 (Hbk)
ISBN 978-3-7258-7217-6 (PDF)
https://doi.org/10.3390/books978-3-7258-7217-6

Contents

materials

MDPI

Review

Energy Conversion Strategies for Wind Energy System: Electrical, Mechanical and Material Aspects

Anudipta Chaudhuri [1], Rajkanya Datta [1], Muthuselvan Praveen Kumar [1], João Paulo Davim [2] and Sumit Pramanik [1,*]

1 Functional and Biomaterials Engineering Lab, Department of Mechanical Engineering, Faculty of Engineering and Technology, SRM Institute of Science and Technology, Kattankulathur, Chennai 603203, India; anudiptac@gmail.com (A.C.); rajkanyadatta99@gmail.com (R.D.); mpk16199@gmail.com (M.P.K.)

2 Department of Mechanical Engineering, Campus of Santiago, University of Aveiro, 3810-193 Aveiro, Portugal; pdavim@ua.pt

* Correspondence: sumitprs@srmist.edu.in or prsumit@gmail.com; Tel.: +91-44-2741-7863

Abstract: Currently, about 22% of global electricity is being supplemented by different renewable sources. Wind energy is one of the most abundant forms of renewable energy available in the atmospheric environment due to different air-currents spread over the troposphere and stratosphere. The demand of modern wind energy conversion system (WECS) has increased to achieve a suitable alternate renewable energy source. In this paper, after a brief introduction, the classification of WECS is reviewed with attractive illustrations. The various mechanical materials and electrical components of WECS are discussed. The flow of power in WECS and its control strategies are also been described. The wind energy conversion is carried out with a suitable controlling mechanism for power grid integration. A maximum power-point tracking controller is an effective controlling method to extract the maximum possible power from the turbines. The present trends in WECS and the scope for improvement and future prospects are discussed. The materials used for both the blade and generator have been found to be key elements of wind turbines. Recycling of the polymer matrix composite materials are found to be a great threat to wind power plants, as well as to their supply chain industries.

Keywords: renewable energy; wind speed; wind turbine; power; power signal; torque; composite

1. Introduction

The demand of energy over the globe is unremittingly growing owing to the growing of population, energy consumption per capita and huge industrialization. The availability of conventional fossil fuels is limited [1]. Thus, energy crises in the future will be increasing more. Therefore, the use of renewable energy sources, such as solar [2], wind [3], biogas [4], ocean [5], tidal [6], and so on is emerging at tremendous rate. Among these, wind energy is one of the most promising sources. Wind energy has been used for a multitude of purposes for a millennium across the globe. From the use of sails in boats and rafts, to wind driven mills for grinding grains, wind energy has found its use throughout history [7].

1.1. A Brief History of Turbine Energy Conversion Technology

The early usage of turbines was found in the form of a carousel, like the vertical axis turbines in Persia used for pumping water and grinding food grains in very small scale, around the 10th century. Horizontal axis turbines were adopted in England, which had to be manually oriented according to the direction of the wind currents, thus giving it the name post mill [8]. However, experiments on large wind machines started during the years of World War II with the introduction of a turbine that was capable of generating approximately 1.25 MW, making way for modern day turbine and grid systems [9]. Thus, the demand of modern wind energy conversion systems has been realized. By 2012, the global power production from renewable resources exceeded 1470 GW, within which 282 GW was produced in grids powered by wind farms [10]. Energy consumption is the backbone necessary to meet the daily requirements of civilization. Starting from burning wood or charcoal as a means of thermal energy, the heater systems are currently used from homes to heavy duty boilers or furnaces in metal and iron smelting industries. As a by-product, the harmful effects of these non-renewable and carbon-based fossil fuels are multiple. Introduction of cleaner and renewable sources of energy such as tidal, solar, hydro, geothermal and wind have brought about immense changes in the realm of energy systems. Presently, about 22.5% or 0.225 fraction of global electricity is being supplemented by different renewable sources [11]. From 2012 to 2014 the global wind power capacity plunged from about 282 to 370 GW, which composes almost 0.4 fraction of the total renewal electricity, generated worldwide [12]. A recent study also reported that, the energy generated by renewable energy added to the global electricity about 70% or 0.7 fraction of the net capacity by 2017. The wind electricity alone surged by 52 GW, reaching to 539 GW in 2017 [11].

Wind energy has been the most abundant form of energy available in the atmospheric environment with varying air currents spread over the troposphere and stratosphere. Wind energy has been used for a multitude of purposes for thousands of years across the globe. From the use of sails in boats and rafts, to wind driven mills for grinding grains, wind energy has found its use throughout history. Wind energy has been a staple for transportation in oceans until as recently as the 19th century. The early usage of turbines was found in the form of carousel like vertical axis turbines in Persia for pumping water and grinding food grains in a very small scale, around the 10th century [13]. Horizontal axis turbines were adopted in England, which had to be manually oriented according to the direction of the wind currents, thus giving it the name post mill. Large scale harnessing of wind energy has been implemented since the 17th century in Europe [14]. These have paved the way for the more effective and useful utilization of wind in the 21st century, with modernization in design and optimization enabling us to store energy in the form of electricity, using wind turbines [15]. The wind energy conversion system (WECS) is an integrated system comprising of wind turbines, generators, mechanisms for control and an integrating method. The turbines are responsible for converting the power of wind into mechanical power, thus harnessing the kinetic energy from the wind. The efficiency of the turbines is a function of shaft power recorded at the rotor shaft and the power available in the flow of the wind stream [16]. The mechanical components are used chiefly for the conversion to mechanical power and transmission to electrical components of the system, as well as providing support to the moving components. The amount of power that can be strained from the wind current depends upon a multitude of factors such as density of wind, shape of turbine rotor, blade measurements, wind velocity, and so on. The high power WECS are low speed, high torque at the turbine. These turbines produce power of more than 1 MW where the speed of the turbine shaft is in the range of six to twenty rpm. However, the generator shaft is that of high speed and low torque, and thus, to couple it with the rotor shaft a gearbox is added [17]. For an effective reduction in torque, a multiple stage gearbox is used in high power grid-connected WECS. A planetary gear assembly ensures even more effective reduction phase. The electrical components of the WECS mainly comprise of a generator, a controller system coupled with anemometer to detect the speed of wind available to control the operation of the turbine.

1.2. History of Wind Turbine Technology

The invention of lead acid batteries by Camille Faure in 1881 provided a viable option for researchers to advance their studies for harnessing wind energy through turbines, since lead acid batteries were lighter and smaller. In the following years, prominent researchers such as Professor James Blyth, and Charles Brush successfully built some forms of horizontal axis turbines that were capable of charging up the accumulators available and illuminate a single mansion or real estate [18]. In the 1890s, a significant surge of wind turbine technology in Denmark resulted in a peak power capacity of 30 MW harnessed by more than 2500 windmills across the country [19,20]. This figure metamorphosed to each windmill to produce approximately 25 kW by the end of the first decade of 1900s, thus doubling the efficiency of turbine technology [21]. During this period, the number of blades on the turbines reduced significantly as lesser blades rotating faster proved to be more efficient than more blades rotating slowly. This was found by Danish researcher Poul La Cour. Eventually, research brought back the design of the vertical axis turbine system. The Darrieus wind turbine was the first among the vertical axis wind turbines (VAWTs) patented by Georges Jean Marie Darrieus, a French aeronautical engineer [22]. However, experiments on large wind machines started during the years of the Second World War with the introduction of a turbine that was capable of generating approximately 1.25 MW [18]. This build was concluded by an American engineer Palmer Cosslett Putnam along with the Morgan Smith Company (Ashaway, RI, USA) that manufactured hydroelectric turbines. This turbine was the first one to be connected to an electricity distribution grid due to its huge power production capacity. Research on wind energy conversion slowed down during the middle of twentieth century due to the feeling that there was less need for wind energy since exhaustible resources for energy were available on a large scale. Not until the end of the century was there much progress observed. In 1978, the first multi-megawatt turbine was created by the Tvind School [23]. The turbine had a novel wing construction, which was made with assistance from German aeronautic specialists [24]. Not only did this prove to be a landmark in the technological advancement of turbines, but it also stands as the oldest functioning wind turbine in history. National Aeronautics and Space Administration (NASA, Washington, USA) pioneered the turbine technologies with their research that included a series of turbines. This knowledge is incorporated into the design of the most recent turbines, such as composite blade materials, steel tube towers, variable-speed generators mounted on the tower along with dynamic components such as aerodynamics of the blades in horizontal axis turbines, pitch and span control mechanisms [25]. A large-scale variable drive train wind turbine capable of creating 3.2 MW was patented by NASA [26]. It was a two-bladed structure. Over the following decades, research was undertaken to improve the aerodynamics of rotor and turbines that have boosted the efficiency of wind energy systems. Uniform wind has yields of around 20–30% efficiency in modern day turbines, and more and more wind farms are being set up, increasing the global wind power capacity annually. The year 2012 marks a landmark with the global wind capacity clocking in at almost 300 GW worldwide, which balanced the cost of wind power to be $0.06/kWh, and is less than one-third the cost of electricity in the 1980s [27].

1.3. Review Outlook

The demand for the modern wind energy conversion system (WECS) has increased to achieve a suitable alternate renewable energy source. In this review paper, the various mechanical and electrical components of WECS are discussed. The flow of power in WECS and its control strategies are also described. We elaborate on wind energy conversion with a suitable controlling mechanism for power grid integration. A maximum power-point tracking controller has been found as an effective controlling method to extract the maximum possible power from the turbines. The present trends in WECS and the scope for improvement and future prospects are also discussed. The typical materials used for the major components are discussed with neat illustrations. This review article will make the researchers' job of selecting the materials best suited for the blade and generator easier.

2. Classification of Wind Energy Conversion Systems (WECSs)

The WECSs fall into various categories depending on different factors. The most common classification of WECS is the one based upon the axis of rotation of the system or rotational axis, turbine, power control, and rotational speed control criteria.

2.1. Rotational Axis

There are two basic axes of rotation possible for a turbine with respect to the ground, such as the horizontal axis wind turbine (HAWT), which rotates in an axis parallel to the ground, and the vertical axis wind turbine (VAWT), which rotates in an axis perpendicular to the ground. The function of both types of turbine systems is to produce kinetic energy through wind pressure generated by the turbine and convert this into electrical energy.

a. Horizontal Axis Wind Turbines: The shaft of the turbine and the shaft of the generator are placed parallel to the plane of the ground and mounted to a certain height to face the required wind speed depending on the geography. The elevated height of the turbine also facilitates the turbine to rotate with enough clearance to the ground. The mechanical components of the turbine are enclosed in a housing, which is shaped aerodynamically called a nacelle. The nacelle in this case is directly behind the rotor. According to the placement of the nacelle with respect to the turbine and tower, there are upwind and downwind HAWTs as shown in Figure 1 [28,29]. Upwind HAWTs have the turbine facing the direction from which the wind is blowing. The direction is adjusted automatically by the use of rudders. The downwind HAWTs have the turbine facing away from the direction of the wind flow. This type of HAWT faces the problem of vibrations in the blades due to the wind interacting with the nacelle first, making the wind turbulent. Commercial turbines incorporate the HAWT design mostly with three blades. However, two blade designs are also used on a very small scale. HAWTs demand high structural strength since the turbine, gearbox, generator, and so on are placed high above the ground.

b. Vertical Axis Wind Turbines: This type of turbine has the turbine mounted on the shaft, which doubles up as the tower of the turbine structure and the rotor's blades are vertically placed aerofoil sections. However, various designs of VAWTs are available, like Darrious and Savonious rotor turbines [30]. The installation and maintenance of these types of turbine are relatively cheaper and easier than other types. Since the turbine blades do not have to clear the distance between the ground and the hub of the rotor, the length of the tower can be kept short, which also helps to reduce the vibration of the shaft due to the rotating mass. However, the shaft does not need to hold the weight of the entire gearbox and control mechanisms as those components are placed at the base of the shaft near the ground. VAWTs are subjected to varying wind loads, differing with the height of the blade. Higher wind loads are experienced at the top of the turbine than the loads experienced at the bottom, rendering lower energy conversion efficiency. Thus, their uses for high-power application are restricted commercially.

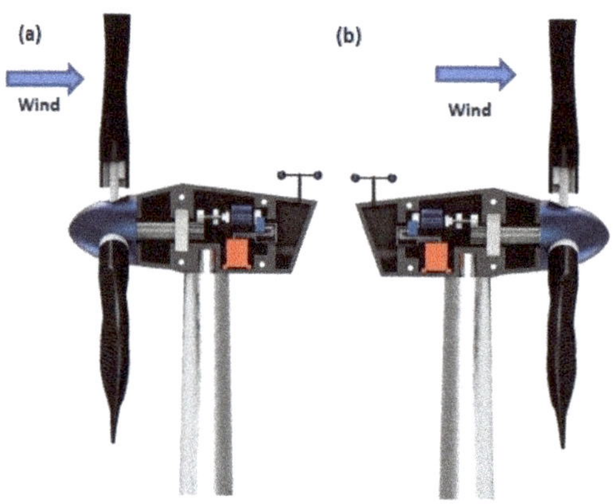

Figure 1. (**a**) Upwind HAWT (when wind is coming from the front side of the turbine blades); (**b**) Downwind HAWT (when wind is coming from the back side of the turbine blades).

2.2. Turbines

Turbines are also categorized based on their electrical output. The output power determines the size of the wind farm. Based on this, the prevailing turbine system technologies can be divided into three categories which are:

a. Low Power turbines: These are turbine systems capable of producing on average a maximum of 30 kW. These systems find use in remote locations for household electricity demands and charging up of batteries. They are also used in the case of an emergency to reduce dependence on primary power sources;
b. Medium Power Turbines: Turbines pushing out in the spectrum of 30 kW to 300 kW fall in this segment. However, they are mostly used to power up the homes in small localities. They are used along with other sources of renewable energy or other power storage devices;
c. High Power Turbines: These refer to the systems where a large production of electricity is undertaken. These are incorporated in the large size wind farms that are associated to the power grids responsible for transporting the energy throughout cities.

2.3. Power Control

Power control ability plays an important part in the design of turbines, since they are completely dependent on wind flow for generating power. As mentioned earlier, each turbine is rated for producing certain power. This power gets overshot when the turbine is subjected to stronger winds than they were designed to withstand. While, this produces more power, there are possibilities of mechanical failures due to increased vibrations and resulting unbalanced loads. Thus, power production is restricted under increased wind speeds, which can be achieved by three controlling methods, namely passive stall, active stall and pitch controllers. The other control methods include grid side controller and phase lock loop controller. Besides, a crow bar circuit is also used to act as protection for the whole electrical circuitry [31]. To obtain optimal maximum energy from the wind resource, different optimization techniques are utilized. These are done by modeling the different components of the WECS and simulating the results.

a. A passive stall controller of WECS controls the input power of the wind by virtue of the aerodynamics of the blades and does not include sensors or actuators to stop the motion of the blades. In such WECS, the blades are fixed so that there is no relative motion with the hub, generally incorporating bolted or riveted connection. The design poses a stall when the wind flowing through the rotor exceeds a certain speed. This takes place since the first blade that comes in contact with the wind turns the flow into a turbulent one and the other blades continue to rotate within the wake of the first blade, thus significantly reducing the lift force experienced by those blades, stalling the turbine. Mostly used in fixed type fixed speed WECS, these are the cheapest and simpler designs than other types of WECS;
b. An active stall control system is similar to the passive system; however, it uses sensors to predict the wind speeds and changes the pitch of the blades by means of motors or hydraulic actuators to interfere with the wind in a certain manner that induces a high turbulence and thus creating a stall. Active stall control systems are observed in high power turbines;
c. Another type of WECS in this category uses similar mechanism as the active stall-controlled design, but the action is slightly different. Unlike the active stall controller, the pitch controller changes the pitch of the blades so that they don't produce lift by reducing interaction with the flow of wind. This is different than the previous type which depended on increased turbulence, which still utilizes aerodynamics of the blade. This control system produces faster response times and thus, it is used extensively at present in the high-power turbine systems.

2.4. Rotational Speed Control Criteria

The criteria of rotational speed control classify the grid connected with wind energy conversion devices into fixed- and variable- speed systems. Here, tip speed ratio is a required concept to explain the classification effectively. In a wind turbine, the tip speed ratio is defined as the ratio of tangential velocity of the blade-tip to that of the wind driving speed of the turbine. When the turbine of radius (r_T) is rotating at an angular velocity (ω_T rad/s) and the velocity of wind (v_w) acting upon it, the tip speed ratio (λ) can be expressed by Equation (1) [28], which identifies the relation between the λ and the design parameters of the turbine.

$$\lambda = \frac{Tangential\ speed\ of\ blade}{Speed\ of\ Wind} = \frac{\omega_T\, r_T}{v_w} \tag{1}$$

a. Fixed Speed WECS: This type of turbine system composes the first generation of commercially set-up grid connected turbines. They are well proven over the ages and have reliable and simple construction. However, the simple nature of these turbines makes them operable only at a certain rated wind velocity, thus the name fixed-speed WECS. Winds at a faster velocity reduce the power coefficient (C_p) of the turbine. The C_p of a turbine is a unitless constant, particular to the design of the turbine. It is a ratio of power captured by the turbine and the available power in the wind due to the flow. The mathematical representation of the power coefficient is expressed in Equation (2) [28,32].

$$C_p = \frac{Mechanical\ Power\ (P)}{Wind\ Power\ (P_w)} = \frac{P}{\frac{1}{2}\rho A v_w{}^3} \tag{2}$$

where, P_w is the power of wind flowing at v_w and can be expressed by $P_w = \frac{1}{2}\rho A v_w{}^3$.

Now, the mechanical power ($P = \tau \cdot \omega$) captured by a rotating object is defined as a product of torque (τ) and angular velocity (ω). Here, the C_p can also be expressed as Equation (3).

$$C_p = \frac{(\tau \cdot \omega)}{\frac{1}{2}\rho A v_w{}^3} \tag{3}$$

Torque applied by wind is a function of density (ρ) and velocity (v_w) of wind, area of turbine rotor (A) and diameter of transmission gears (d), and can be expressed by Equation (4). Therefore, the power coefficient (C_p) again can be given by Equation (5) [28].

$$\tau = \frac{1}{2}\rho A v_w{}^2 d \tag{4}$$

$$C_p = \frac{\frac{1}{2}\rho A v_w{}^2 d \cdot \omega}{\frac{1}{2}\rho A v_w{}^3} = \frac{1}{v_w}(d\omega) \tag{5}$$

The Equation (5) demonstrates that the value of C_p is directly proportional to the ω and inversely to the velocity of wind stream (v_w). This implies that at a constant wind stream velocity, higher turbine speed yields higher coefficient of power. This equation demonstrates that with increasing wind velocity the coefficient of performance decreases. Since power generated by the turbine is a function of C_p, the power produced decreases as well. Thus, the drivetrain experienced an increased mechanical stress, while the power conversion efficiency falls significantly, nullifying the use of fixed speed turbines in commercial purposes.

b. Variable Speed WECS: The variable speed WECS were innovated from the older designs, that is the fixed speed WECS, considering the limitations of the latter, as utilization of wind power in power systems started to grow. Components of WECS are depicted in Figure 2 [33]. The limitations of the older designs were pronounced in places where the grids were not as strong to compensate for loss of power at higher speeds of wind. The development of power electronic convertors paved the way for large scale applications of the variable speed WECS. Power electronics are employed to make a connection between the turbines and the grid. The main advantage obtained by this system was that the rotor speed can be controlled according to the speed of wind. Thus, this system is able to maintain the optimum tip speed ratio by producing even and smooth power throughout the operational range. Since the optimum 'λ' is maintained, the system can thus maintain the designed value of C_p, which indicates maximum aerodynamic efficiency. It is a relatively complex and costly system, however the high yield of the system compensates for these shortcomings. Present wind energy systems use them on a large scale.

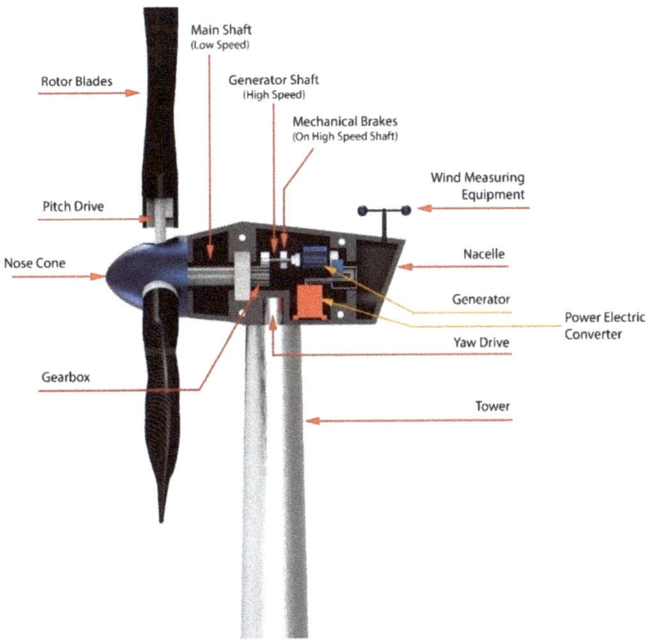

Figure 2. Components of WECS.

3. Components of WECS

The WECS is an integrated system comprised of a symphony between the mechanical and electrical components. Among the various designs of wind energy conversion systems previously discussed, some components bind them together. These systems indeed share lots of components, which are able to convert the kinetic energy of wind into electrical energy in order for it to be stored in the cells efficiently. One of the biggest parts of grid connected WECS is the wind turbine assembly itself, which is the biggest tangible component. The wind turbines are compound mechanical machines built up of several small components responsible for feeding the electrical components. The aforementioned Figures 1 and 2 represent the cross section of a typical horizontal axis wind turbine. A typical classification of the components can be: mechanical components, electrical components, and control system components.

3.1. Mechanical Components

Mechanical parts of a wind turbine have a great role in conversion of the energy. The flow of wind has some energy in it due to their kinetic nature. The mechanical regime caters to introduce a 'work' in the system by allowing a mass to rotate and transmit it to the desired location to interact with the electrical components. The rotor subassembly is the rotating body, which includes rotor blades, rotor bearings, nose cone, hubs, and the pitch drive system.

Some of the major mechanical components that are characteristics of WECS are:

a. Rotor: Rotor assembly is the aggregate of the barely distinctive elements that interact with the wind. The assembly consists of the blades, the hub and the nose cone. The blades are the aerodynamic devices responsible to initiate the drive. In HAWTs, they resemble the propeller of aeroplane, which operates on similar principle. The blade of a typical horizontal axis wind turbine generates lift, which turns the blade during wind flow in the axial direction. The same is not true for a vertical axis wind turbine, which can either be a lift type (which uses the lift forces generated by the moving air to produce the rotation of turbine blades, see Figure 3a) or a drag type (which uses the drag force of wind to produce the rotation motion of blades, see Figure 3b). A lift type vertical axis wind (VAW) is capable of having a higher aerodynamic coefficient, and thus are used more, however, they are mostly incapable of self-start since the torque produced is lower. Over the years, research is being conducted to find the ideal materials for production of the rotor blades. Initially, they were made from heavier metals, which were resistant to environmental conditions such as high turbulent wind, however were rendered not helpful since metals tend to have higher density and thus higher mass. Concentration of mass away from the center catered to a higher rotational moment of inertia, thus diminishing the acceleration of the rotor. With a higher mass concentration away from the center, the moment of inertia of the system gets high. As a result, the acceleration gets lower, which in turn results in less response to wind speed. A massive component would only begin to be affected by winds that can exert a minimum value of force, effectively increasing the cut-in speed. In addition, another downside would be the cut-out scenario, which would require more energy dissipation through braking to keep the rotation below a certain speed, in case of high wind speeds. This turned the focus towards composites that provide rigidity while not contributing to mass. The hub of the rotor facilitates mechanical support for the rotor assembly to stay upright while the latter is rotated. It houses the bearing to provide relative motion between the fixed support and the shaft. The nose cone is integrated with the hub, which primarily directs the wind interacting with the center of the rotor towards any of the three blades. Traditional rotor designs mandated a rigid connection between the hub and the blades. The modern designs have deviated from it to include a pitch control mechanism responsible for rotating the blades about its own axis to reduce lift and control the speed of the turbine [34].
b. The Main Shaft: The main shaft is a solid extrude made from any of forged high carbon steel or cast iron/steel that attaches the gearbox to the rotor hub. At present, the megawatt size wind turbines rotate very slowly and the torque is high. Thus, the main shaft is also called low-speed shaft. For example in a 5 MW power turbine, the input shaft speed is about 12 rpm while the torque is in the range of 4000 kNm [35]. This torque value can be considered to be high from a structural point of view, as a torque transmission of 4000 kNm would require approximately a 9 m cross section of EN316 type steel.
c. Gearbox: The main shaft rotates slowly at a very high torque. These rotational parameters are unusable by the generator, which require high speed low torque input. The gearbox is thus included between the main shaft and generator. Modern megawatt wind turbines operate on an overall torque reduction of 60 to 120 times [36]. Research, by Nejad et al., demonstrated the use of a gearbox with an overall gear ratio of 1:96.354 [35]. For such high reduction, multi-stage gearboxes are important where one or more stage incorporates a planetary gear system, since they are capable of huge gear ratios in a small assembly space. Since gearboxes are subjected to varying torques as the input shaft of the gearbox is bolted directly to the rotor hub, the gear train experiences high wear and tear.

d. Mechanical Brakes: Brakes are safety equipment installed inside the housing for emergency situations for stopping the turbine during storms or high velocity winds. They are installed directly on to the low torque high-speed generator shaft for efficient braking, rather than on the high torque low speed main shaft. This keeps them from overheating and failing during emergencies by keeping the braking torque low. Modern high-power turbines are fitted with hydraulic or electromechanical actuated brakes (both disc and drum type brake mechanisms). Braking from maximum speed of the generator shaft can cause high wear of the latter and can also cause fire hazards due to increased temperatures. Hence, they are used after proper actuation of pitch and/or yaw drives, which reduce the initial speed of the rotor.

e. Nacelle: It is the enclosure housing most of the mechanical components. In horizontal axis wind turbines, the nacelle is the structure mounted on the tower behind the rotor. The size and shape of the nacelle is influenced greatly by the design of the gearbox and reduction required. This fiberglass construction is constructed to reduce turbulence to enable less vibration through the structure. Manolesos et al. demonstrated the aerodynamic properties of generic wind turbine nacelle designs [37]. Vertical axis wind turbines feature transmission and gearbox assembly at the base of the tower, thus, neither an aerodynamic enclosure nor light weight construction is necessary.

f. Pitch and Yaw Drives: Pitch of turbine blades refer to the rotational degree of freedom about the longitudinal axis of the blade and the yaw is the same for the top of the turbine (including the rotor and nacelle) with respect to the tower. With regard to pitch in the classification of WECS, as we have discussed, every turbine is rotated to function at a certain wind speed and produce a designed power. Since the power generation of a turbine depends on the energy captured from wind during higher velocity of operational winds, the blades are "pitched" to alter the value of aerodynamic efficiency C_p. Each blade is fitted with independent pitch control mechanism for this purpose. Likewise, the yaw drive is employed to align the turbine according to the direction of wind flow in order to obtain maximum efficiency. Discussion about control mechanisms and strategies would be covered further with greater details.

g. Wind Measuring Equipment: As mentioned in regard to the pitch and yaw drives, the turbines operate on complex mechanisms, however the control of those mechanisms is ultimately managed by monitoring environmental conditions. Thus, turbines are further equipped with any kind of mechanical anemometer to measure the wind velocity around the system. An anemometer is a miniscule vertical axis turbine like arrangement with cupped struts, mounted on to the nacelle. Further, wind vanes are utilized in some wind turbines for judging the direction of flow, based on which of the controller signals the yaw drives to align itself to the wind.

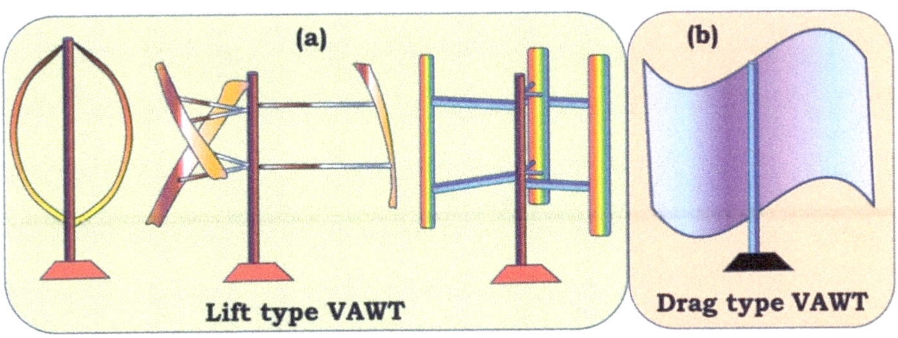

Figure 3. (**a**) Lift type vertical axis wind turbines (VAWTs) and (**b**) Drag type VAWT.

3.2. Electrical Components

Electrical components are separated into two segments, some of which are located inside the nacelle alongside the mechanical components, while others are placed on the ground away from the tower. The chief function of the electrical subsystem is to convert the mechanical energy gathered from the wind as torque and speed into electrical energy. A block diagram of a typical grid-connected WECS is depicted in Figure 4 [28]. The step by step conversion begins with the generator, power electronic converters, transformer, wind farm collection point, and power channels etc.

a. Generator: A wind generator is an electromechanical component responsible for converting rotational motion into electrical power, thus converting rotational kinetic energy into electric potential. A typical AC generator works by moving a conductor loop in a static magnetic field based on the electromagnetic induction law of Faraday. Traditionally various types of generator have been in use with the introduction of a squirrel cage induction generator (SCIG) almost 35 years ago [28,36] along with a wound rotor induction generator (WRIG). Other generators being used are doubly fed induction generators (DFIG), permanent magnet synchronous generator (PMSG), and wound rotor synchronous generator (WRSG). The DFIG based WECSs have the stator present in the generator tied to the grid, while the rotor is linked to power converters [7]. Ideally, generators are optimized to produce power at the rated wind speed, above which control systems start operating to protect the systems from damage and brakes are applied on to the generator shaft [36]. Popular choices of generator amongst WECS are DC, AC Synchronous and AC Induction [36].
b. Power Converter: The output electric parameters from the generator vary widely according to the wind speed. Thus, each turbine would produce different voltages and frequencies of current at a particular instant, which is not possible to be directly coupled to the grid without refining. A power converter thus acts as an interface for connecting the turbine to the grid. The power converter transforms the output voltage from AC to DC using a rectifier and again the rectified DC to AC, having a constant voltage and frequency in subsequent steps using inverter circuit [36,38].
c. Step-up Transformer: The output from the generator in a grid connected MW turbine is rather low, in the range of 400 to 690 volts, which requires it to be stepped up since the grid connects directly to the high-tension lines [39]. The necessity of a step-up transformer can be annihilated by designing the generator and power electronic converter according to the wind farm collection point voltage, however it demands additional costs for medium voltage generator and power converter setup, diminishing the economic benefits of removing a step-up transformer from the circuit [40]. However, studies are being conducted in recent years to eliminate the needs of transformers in medium voltage grids in order to reduce transmission losses incurred from stepping up the voltages required for transmission [39,41].
d. Wind Farm Collection Points or Point of Common Coupling: The point of common coupling (PCC) is a connecting ground of all the turbines of a wind farm. The preferred mode of connection in the wind farms is a parallel connection, which facilitates either maintaining a desired potential difference or having a defined node for connecting more turbines if necessary [41].

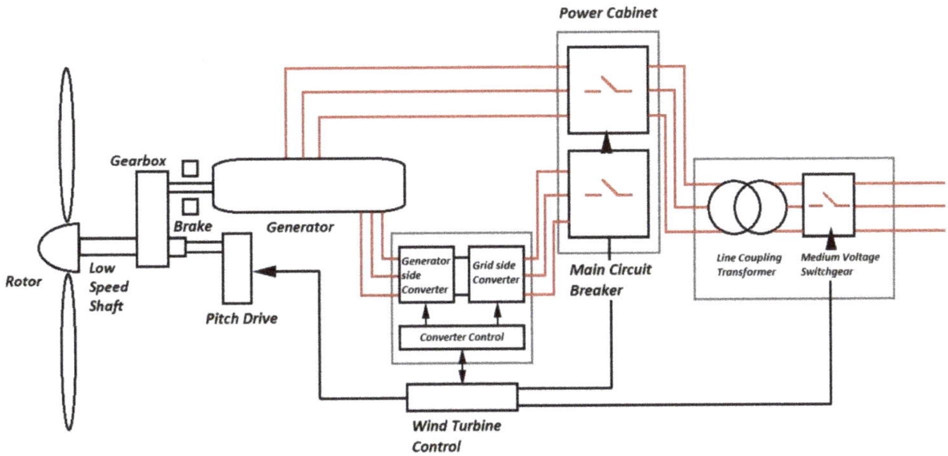

Figure 4. Block Diagram of a typical grid-connected WECS.

4. Flow of Power in a WECS

The WECS is a constant speed or constant frequency device, which is fitted with an induction generator, which is operated by a wind turbine without any controller. The power content in a stream of wind is a function of wind flow velocity. When the speed of wind stream is V, the power (P), which is a rate of change of kinetic energy of wind according to definition can be expressed in Equation (6) [42,43].

$$P = \frac{dE}{dt} \tag{6}$$

Kinetic energy of a system is given as product of mass (m) and square of velocity (V) as given in Equation (7) [12,13].

$$E = \frac{1}{2}mV^2 \tag{7}$$

Inserting the expression of kinetic energy in the Equation (6), we get, $P = \frac{d}{dt}\left(\frac{1}{2}mV^2\right)$. Differentiating the expression with respect to t, $P = \frac{1}{2}\left(V^2\frac{dm}{dt} + 2mV\frac{dV}{dt}\right)$. Since the wind velocity, V is constant, thus, $\frac{dV}{dt} = 0$. Therefore, the relation of power with mass flow rate of the wind ($\dot{m}$) is arrived at from energy in Equation (8) [44].

$$P = \frac{1}{2}\dot{m}V^2 \tag{8}$$

Let the cross-sectional area of the wind column in consideration be A and the air density be ρ. Therefore, the mass flow rate can be written below in Equation (9) [43].

$$\dot{m} = \rho AV \tag{9}$$

Therefore, Equation (8) can be rewritten utilizing the value of mass flow rate from Equation (9). This can be written as given in Equation (10) [44].

$$P = \frac{1}{2}\rho AV^3 \tag{10}$$

The Equation (10) represents the power content in a steam of wind only due to its flow and velocity. However, similar to Carnot heat engine, a turbine system cannot theoretically extract the whole power from the flow. This is governed by Betz Equation, which deals with upstream velocity of wind, V_1 and downstream velocity of wind, V_2 and the maximum extractible power is called the Betz limit. The wind turbine cannot extract the whole available power due to breaking of wind velocity from V_1 to V_2, allowing a continuation of wind regime [44]. The estimation of Betz limit assumes the turbine to be ideal, that is:

a. The turbine has no hub and has an infinite number of blades with no drag from any of the blades (ideally zero drag);
b. The flow is non-compressible, that is, the density of the air remains constant throughout;
c. Thrust experienced by the rotor is uniform throughout.

Betz equation further considers the flow to be axial to the turbine. The ideal turbine is in a non-rotational form at the time of placing it in the wind atmosphere. The cross-sectional area of swept by the blade is A while the cross-sectional area upstream and downstream are designated the values S_1 and S_2, respectively.

The speed of wind interacting with the rotor is taken to be V while the velocity of wind upwind from the rotor is designated to be V_1 and the downstream velocity is designated to be V_2. By definition, the turbine extracts mechanical energy from the wind by reducing its kinetic energy from the upstream to the downstream. Therefore, V_1 would be greater than V_2 and subsequently, the cross-sectional area enhances from the upstream to the downstream, and hence, S_1 would be lesser than S_2. Since the stream of air is considered incompressible from the listed assumptions, the mass flow rate remains constant throughout.

The force experienced by the rotor of the wind (F) can be obtained by applying Euler's theorem [45] i.e.,

$$F = ma = m\frac{dV}{dt} = \dot{m}\Delta V$$

Power flux is an essential fluid parameter, which defines the rate of energy flowing through a given cross-sectional area. By definition, power flux ($P' = P/A$) is the rate of energy flow per unit area. The cross-sectional area is pre-defined to be A, and the rate of energy flow is effectively the power flow, P.

Using general value of P from Equation (9), we can arrive at the relation given by Equation (11).

$$P' = \frac{\frac{1}{2}\rho A V^3}{A} = \frac{1}{2}\rho V^3 \tag{11}$$

The kinetic power content in the air upstream to the rotor is obtained by replacing the velocity term with upstream velocity V_1 in over a cross-sectional area A, which is demonstrated in Equation (12).

$$W = \frac{1}{2}\rho A {V_1}^3 \tag{12}$$

We can introduce a dimensionless number obtained as the ratio of the net power extractable from wind, to the total power content of the same wind stream. This designates the coefficient of performance or efficiency of extracting power. The coefficient of performance represented by C_p is also known as the aerodynamic efficiency since turbines rotate by the action of aerodynamic properties of the blades. It signifies the available power percentage extracted by a turbine. The C_p value can be expressed in terms of P and W and we get a simplified relation as expressed in Equation (13) [42,44].

$$C_p = \frac{P}{W} = \frac{\frac{1}{4}\rho A {V_1}^3 (1+b)(1-b^2)}{\frac{1}{2}\rho A {V_1}^3} = \frac{1}{2}(1+b)\left(1-b^2\right) \tag{13}$$

Equation (13) gives a sensible relation between the coefficient performance and velocities of wind before and after hitting the turbine. Setting the value of b as zero in Equation (13), we get C_p to be 0.5 and setting b as 1, we get C_p to be zero. Thus, the nature of the curve needs to be investigated in this domain. Moreover, putting $b = 1$ gives $V_1 = V_2$, which represents no kinetic energy being extracted from the wind and $b = 0$ gives $V_2 = 0$, which represents no wind flow beyond the turbine. Between the points $b = 0$ and $b = 1$, a point of zero slope would give the maximum value of C_p. Therefore, differentiating C_p with respect to b when $\frac{dC_p}{db} = 0$, will get a relation as derived in Equation (14) [46].

$$\frac{dC_p}{db} = \frac{1}{2}\frac{d}{db}\left[(1+b)\left(1-b^2\right)\right] \ or, \ 0 = \frac{1}{2}(1-3b)(1+b) \quad (14)$$

The Equation (14) is derived to find a suitable solution. The equation holds for two values of $b = -1$ or $b = \frac{1}{3}$. When $b = -1$, $\frac{V_2}{V_1} = -1$ or $V_2 = V_1$.

This gives a trivial solution for the Equation (11) since the direction of the wind cannot be reversed and the range of V_2 is in positive values of V_1. Therefore, this solution should be ignored when $b = \frac{1}{3}$, $V_2 = \frac{1}{3}V_1$. Thus, the velocity is decreased by two-thirds and the initial stream velocity for maximum achievable C_p. This has been validated experimentally in a wind tunnel and mathematically modeled in line with many other studies [18]. Therefore, the value of $b = \frac{1}{3}$ can be used in Equation (13) to find the value of C_p, which is expressed in Equation (15) [46].

$$C_p = \frac{1}{2}\left(1+\frac{1}{3}\right)\left[1-\left(\frac{1}{3}\right)^2\right] \cong 0.59 \ or, C_p = 0.59 \quad (15)$$

This value of C_p, obtained in Equation (15), indicates that with highest possible design optimization, the efficiency is still 59% of the available wind energy. Therefore, the Betz coefficient value (i.e., $16/27 \approx 0.593$) indicates that a wind turbine can extract, at most, 59.3% of the energy in an undisturbed wind stream [44]. It is thus the target of modern manufacturers to extract 100% of this extractable value. Recent advancements in technologies have been able to push this value to about 75% to 80%, which is around 45% to 48% of the available wind energy in the free stream [44,46,47].

5. Control Strategies Employed in WECS

As discussed earlier in Section 2.3, the control systems are useful for a turbine for its safe operation. However, to drive those systems, strategies are utilized so that the turbine can generate optimal power for various atmospheric conditions while increasing the life span of the structure by controlling the loading conditions. The strategies determine the operational parameters and change the configuration such as pitch angle, yaw angle, etc. The grid connected WECS incorporates some of the control strategies that will be discussed in this paper. The controls associated to modern day wind energy conversion systems are broken down into six levels of controls, which are chronologically numbered. The first level, or the Control Level 1, deals with slow changing parameters while the Control Level 6 handles the fast-changing parameters. The Level 1 is responsible for imposing power commands to other control levels. The loops in the control hierarchy also monitor abrupt functions of the WECS.

Control in Level 1: Transmission system and distribution system operator (TSO/DSO) controls

Control in Level 2: Wind farm centralized control

Control in Level 3: Wind turbine centralized control

This level can be broken down into mechanical and electrical controls.

Mechanical Controls:

a. Pitch control
b. Yaw control

Electrical Control:

a. Reactive power generation (RPG)
b. Fault ride-through (FRT)

Electro-Mechanical Controls:

a. Damping control
b. Ancillary services

Control in Level 4: Grid connection (GC) and maximum power point tracking (MPPT) control

a. Integration and synchronization of grid
b. MPPT control

Controls in Level 5: Generator and control of grid

a. Speed, torque and power control of generator
b. Voltage oriented control of grid

Controls in Level 6: Power control converter

a. Current control of generator
b. DC chopper control
c. Current control of grid

The whole operation of control works on a system of feedback. In case of faults in the grid, the fault ride-through control of control level 3 generates a fault enable signal, signifying a positive identification for a fault, which is responded to by coordinating for better control performance by mechanical and electrical controls included in control levels 1 to 6. Feedback signals in a traditional wind energy conversion system consist of voltage and current of grid, angular speed along with voltage and current of generator, DC link voltage conditioner, a rotor positioning with respect to mean position and wind speed from the anemometer mounted on the nacelle [48].

Level 1: Transmission system operators or distribution system operators supervisory control.

The output from the turbine or an entire wind farm is not uniform; rather it varies with undulations in wind, which cannot be controlled. The coupling of this unstable energy waveform into the existing pool of energy causes reliability issues and the excessive penetration of a large scale WECS into the existing power system. The WECSs are linked to the TSO/DSO dispatch centers and keeps them updated about the active as well as reactive power generation statuses on a continuous basis. The TSO/DSO sends active and reactive signals based on the received status signals to the respective wind farm setup. Over the years, the large-scale wind farms have changed their operation from being passive sources of power to active ones through constant efforts made by the TSO/DSOs.

Level 2: Wind farm centralized control.

This acts as a communicative medium between the wind farm and the level 6 control. Every turbine in a wind farm is connected to the centralized hub through communication lines. Through it, the status of the WECS is shared to the TSOs and DSOs. The data acquisition and supervisory control systems are responsible for monitoring the operation of the wind farms. The active and reactive power requirements are defined for each specific wind turbine in a farm by the centralized control of that farm, which are transmitted chronologically as per the number of turbines it is in the farm. Since the control is highly reliant on feedback channels, the static compensators of the wind farms, including static synchronous compensator (STATCOM) and static var compensator (SVC), are initiated to meet the reactive power generation (RPG), when the centralized control realizes it is unable to do so normally. Besides all the operations required to meet the references imposed by the higher level, i.e., level 6, the centralized control also checks the aerodynamic interactions.

Level 3: Wind Turbine Centralized Control.

The level 4 of the control system is a cluster of electrical, mechanical and electro-mechanical systems as we have established in the classification. The pitch and yaw control mechanisms are responsible for changing the physical operating parameters and come in mechanical controls, whereas RPG and FRT are responsible for signal modifications, establishing the electrical controls. The third control includes damping, which can be both mechanical as well as electrical. Mechanical damping is necessary to protect mechanical components from torsional vibrations in the drivetrain and resonances in the tower caused by wind, whereas electrical damping provides protection to electrical counterparts by damping electrical sub-synchronous resonance in grids.

The ancillary service acts as an electrical back-up for the mechanical control system and includes the power storage module as well as an uninterrupted power supply (UPS). The energy storage module collects power from idle spinning of the rotors without wind, i.e., rotation due to inertia after wind has fallen below cut-in velocity [49]. As mentioned previously, the MW—powered wind turbines equipped with pitch control mechanism can change the rotation of the blades about their vertical axis. A proportional integral (PI) controller is dominantly uses a thyristor and rectifier kept in between the generator and resistive load. The PI controller works by calculating the integral error to the reference maximum power [32,48]. The PI calculates the exact value of the pitch angle, when the wind speed exceeds the rated wind speed, so that the output power from the generator is limited to its rated output value. However, at wind speeds below the rated one, the pitch angle is kept at zero for maximum possible energy extraction [32,48].

Level 4: Grid connection and MPPT control.

MPPT signifies the control loop that is associated with extracting the peak power at all times. The loop works by using a suitable algorithm on control to adjust the phase between the grid current and grid voltage for it to correspond with the grid power factor requirement. Synchronization of grid parameters is possible by different methods such as zero-crossing, grid voltage filtering and phase-locked loop. Among them, the phase-locked loop is preferred due to the robustness of the controller against the voltage harmonics and simplicity of the platform [49].

With the levels defined, we can now move on to discuss the strategies undertaken by the WECS. Depending on the type of generator used, the control strategies can be broken down as the generator based WECS. The DFIG exhibits the salient feature of variable operational speed due to the integration of the variable resistor in the circuit. Presently, the widest used generator in WECS is DFIGs. The following controls are therefore in the context of DFIG based WECS.

a. Optimal TSR (OTSR) Control: The maximum peak point tracking as discussed previously, utilizes algorithms to estimate reference operational speed (ω_m^*) for the prevalent wind speed (v_w) considering other physical parameters of the turbine such as pitch angle, as well as feedback signals for required power factor. The ω_m^* is tuned based on the wind velocity, such that the optimum tip speed ratio (λ_T^{op}) can be churned. The adjustment of the speed is done based on the expression as given in Equation (16):

$$\omega_m^* = \omega_T^* r_{gb} = \left(\frac{\lambda_T^{op} r_{gb}}{r_T} \right) v_w \quad (16)$$

where, ω_T^* is the angular velocity of the rotor; r_{gb} is the gear ratio employed in the gearbox and r_T is the radius of the rotor, in compliance with Equation (1). The control levels 1 and 2 operate on feedback from the generator. The feedback from generator is a function of the rotational speed of the generator. The control level checks for the error between the obtained value and the reference velocity for every instant and attempts to minimize this error. As this control technique does not involve memory, since the data are processed and compared in every instant, the complexity of the system is lowered. This MPPT control strategy is widely in use in WECS industry, being the simplest among all MPPT control strategies [50,51]. However, proper measurement of wind speed is an important criterion, which demands the use of expensive equipment. Ultrasonic sensors are promising in this regard due to their ability to furnish reliable and accurate wind information.

b. Turbine Power Curve (TPC) based Control: Similar to the previous control method, MPPT strategy utilizes the wind speed for controlling the turbine speed. In this process, the power produced based on wind speed curve is referenced to the power curve provided by the manufacturer of the wind turbine, which is initially tested in controlled wind speeds, which provides the output power curve. Each wind speed and corresponding power, P_0^* output is stored in a memory space. During operation at a particular wind speed, the power produced by the generator can be compared with the data point at a same wind speed from memory. Here, the output signal from the generator is in terms of instantaneous voltage, V_0 and current, I_0. Power is obtained as $P_0 = V_0 \cdot I_0$. The control levels 1 and 2 attempt to reduce the error between the power values P_0^* and P_0 by response signals, which closely resemble the ones used in "optimal tip speed ratio control" [52].

c. Optimal Torque (OT) Control: The previous methods rely completely on the speed of wind to govern the control of the operational parameters. Therefore, even with the implementation of expensive wind speed measurement equipment, the input information is still incomplete. Thus, for increased reliability, other MPPT control techniques also utilize the rotational speed of the generator rotor. In this, the speed of the generator is used to calculate the reference electromagnetic torque T_e^*.

We can conveniently rearrange the expression in Equation (2) to get the power produced by the rotor and by substituting the value of v_w into this equation from Equation (1), we get the mechanical power as Equation (17):

$$P = \frac{1}{2}\rho A {v_w}^3 C_p = \frac{1}{2}\rho A \frac{{\omega_T}^3 {r_T}^3}{{\lambda_T}^3} C_p \tag{17}$$

The turbine torque (T_T) and the turbine speed (ω_T) can be stated by Equation (18) and Equation (19), respectively:

$$T_T = \frac{P_T}{\omega_T} \tag{18}$$

$$\omega_T = \frac{\omega_m}{r_{gb}} \tag{19}$$

where, P_T = Power output from turbine, ω_T = rotational speed of turbine, ω_m = rotational speed of generator, r_{gb} = gear ratio of gearbox.

The torque experienced by the generator from the drivetrain is defined by $T_m = \frac{T_T}{r_{gb}}$. At steady state conditions, the electromagnetic torque (T_e) produced by the generator should be equal to the mechanical torque (T_m) experienced from the drivetrain. Therefore, at T_e = T_m, from the above relations, the torque can be written as a function of angular velocity of the generator by Equation (20):

$$T_e = T_m = \frac{T_T}{r_{gb}} = \frac{P_T}{\omega_T r_{gb}} = \left(\frac{1}{2}\rho A {r_T}^3 \frac{C_p}{{\lambda_T}^3 \cdot {r_{gb}}^3}\right){\omega_m}^2 \tag{20}$$

The reference electromagnetic torque considers optimal coefficient of power and tip speed ratio. Therefore, the reference torque is given by Equation (21):

$$T_e^* = \left(\frac{1}{2}\rho A {r_T}^3 \frac{C_p^{op}}{\lambda_T^{op3} {r_{gb}}^3}\right){\omega_m}^2 = K_{op}{\omega_m}^2 \tag{21}$$

In Equation (21), K_{op} can be treated as a constant for the operation and can be calculated initially with the known design parameters. The torque can be obtained by multiplying this with the angular velocity of the generator. The electromagnetic torque follows a quadratic relation with the speed of the generator evident from the expression. The control in this strategy employs control levels 1 and 2 to equalize the electromagnetic torque to the reference value, by referencing during operation. This method is also used on a large scale in commercial MW-WECS [52].

d. Power Signal Feedback (PSF) Control: Another MPPT control utilizes the input signal in the form of generator speed. This follows the basic principles of Power Curve based control strategy, however instead of monitoring the wind speed, sensors are employed to monitor the speed of the generator. The Equation (16) signifies that the ω_m of a generator is directly proportional to the v_w. Therefore, similar to the power curve used in the wind turbine power curve–based control, this strategy utilizes power versus wind velocity curve to compute the reference power, P_0^*. The control system utilizes a fuzzy logic–based controller that has two inputs, namely, the rotor speed from the generator and the generated power. The estimated output power can be generated corresponding to the conditions. If the input power to the controller is lower than the estimated maximum value, the generator is controlled accordingly to reach more velocity allowing more power to be elicited from the wind [53].

e. Speed Sensor-less (SSL) Control: This MPPT control operates without the use of any kind of speed sensing device. Rather, the wind speed is predicted by autoregressive statistical models of WECSs, which are based on the measured frequency (f_a) of the generator. The predictions implemented regarding the wind speed category based on the basis of time scales are [54]:
    - *Ultra-short-term forecast*: These predictions are regarded for wind conditions about an hour ahead.
    - *Short-term forecast*: This is applicable for forecasts several hours ahead.
    - *Medium-term forecast*: The prediction of wind conditions in the time span from a few hours to about a week.
    - *Long-term forecast*: Used for detecting wind conditions more than a week ahead.

The predicted wind speed is used to calculate the reference power, P_0^* similar to the process incorporated in power curve-based control. Since the wind speed is predicted in this format, the method can also be used congruent to optimal tip speed ratio control, however the increased complexity of this control system impedes its use on a large scale. A summary of MPPT control techniques used for DFIG based WECS is illustrated in Table 1 [28,53].

f. Hill Climb Search (HCS) Control: It is a method of maximum peak point tracking control strategy that involves a disturbance in the reference signal. The inputs in this control are obtained in the form of generator speed and dc-link current. The output is a command signal that is needed for extraction of maximum/peak power. A schematic of the hill climb search control is depicted in Figure 5 [53]. As the generator speed increases, the output power is seen to increase while approaching the peak output power, as illustrated in Figure 6 [53]. After the peak power is encountered, an increase in rotor speed tends to decrease the power output. The mechanical power being generated is compared to the maximum extractable power in this method, after which it is decided whether or not it is necessary to increase the generator speed. The speed of the generator is altered by changing the dc-link's current value. Once the WECS control starts, the controller perturbs for every instant by tracking maximum points with every instance of the operating parameters being changed.

Table 1. Summary of MPPT control techniques used for DFIG based WECS.

Control Technique	Input Parameter	Prior Knowledge on Wind Turbine	Memory Needed	Complication of Control	Performance of Control
Optimal tip speed ratio (OTSR)	v_w	Not required	N	↓	++
Wind turbine power curve (WTPC)	v_w	Required	Y	↓↓	++
Rotational speed of generator Optimal torque (OT)	ω_m	Required	N	↓	+
Power signal feedback (PSF)	ω_m	Required	Y	↓↓	+
Secure sockets layer (SSL)	f_a	Required	Y	↓↓↓	+
Hill-climbing search (HCS)	ω_m, i_{dc}	Required	Y	↓↓↓	+

Yes: 'Y', No: 'N', Low: '↓', Medium: '↓↓', High: '↓↓↓', Good: '+', Very good: '++'.

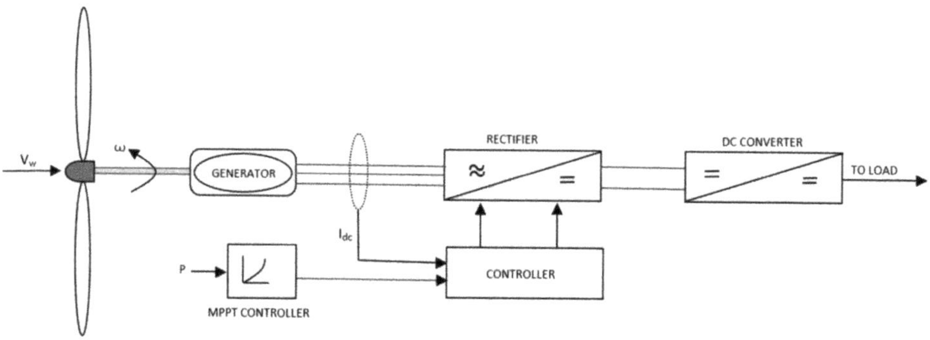

Figure 5. Schematic of Hill Climb Search Control.

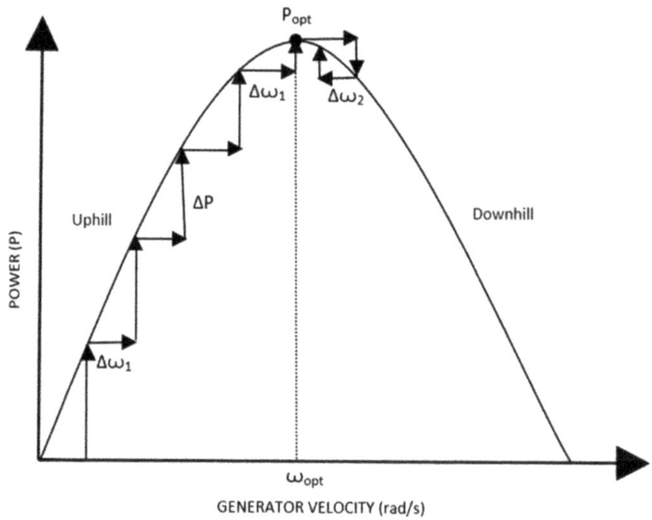

Figure 6. Power curve in a HCS Control.

The control methods discussed above are used precisely in a doubly fed induction generator (DFIG) based WECS. Other examples of WECS architecture, such as the permanent magnet synchronous generator (PMSG) and the squirrel cage induction generator (SCIG) based designs, are substantially more expensive than the DFIG. Moreover, the inverter output filters and electromagnetic interference (EMI) filters used in the previous types of WECS are rated for 1 p.u. of output power, which also contributes to the complexity of designing the filter, making it more costly. The WECS employing DFIG at its core uses back to back converters. The primary advantage thus obtained is the reduced converter rating, which is about 30% of the generator rating; while the speed range of variable speed WECS stays at about 33% of the synchronous speed mechanisms [55].

A PMSG is a staple in the industry for innovating and incorporating new designs, and has attributed to it a higher efficiency alongside greater power density and an achievable smaller diameter of turbine in direct-drive systems. These are used particularly in older generation turbines to keep the complexity lower. Permanent magnet materials of high energy are also available at quite a reasonable price, making these generators desirable. With increasing wind farms, the focus is concentrated on achieving WECS which is reliable, and has lower maintenance cost and noise while being efficient. This converged to direct drive PSMG based WECS. A conclusive design consisting of three MPPT based controls is applied on available wind turbines.

a. Tip Speed Ratio (TSR) Control: Not unlike the optimal TSR control strategy used in case of a DFIG based WECS, the optimal TSR (OTSR) is evaluated with the velocity of wind, v_w as an input parameter. A schematic flow of TSR control is depicted in Figure 7. With the knowledge of the tip speed ratio to be maintained for the given wind speed, the optimal speed of the turbine is to be evaluated by the known relation given by Equation (22) [28,53]:

$$\omega_m^* = \frac{\lambda_T^{op}}{R} v_w \tag{22}$$

where, R is the turbine rotor radius. This speed control command is used further to the control loop responsible for controlling the speed. The PI controller is responsible for processing the command and to limit the speed of the turbine rotor to the desired maximum/peak value of rotor speed obtained by the evaluation.

b. Power Signal Feedback (PSF) Control: The power signal feedback system of control utilizes the information about the rotor speed as input of the controller and power produced as a feedback, directly from the generator. In this system of control, the reference power is evaluated through the algorithm, which is then compared to the produced value of power obtained through feedback loop.

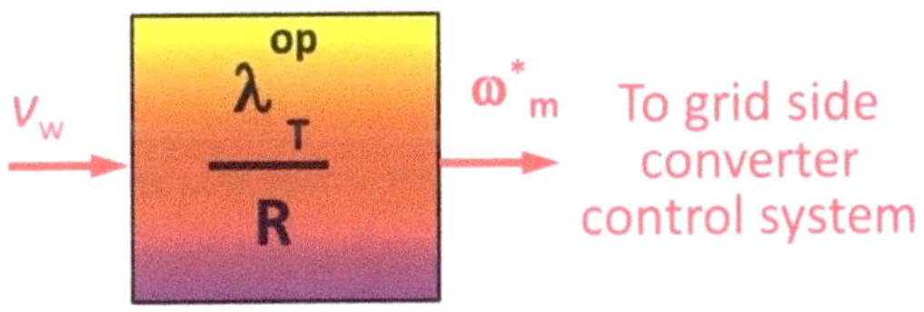

Figure 7. Tip Speed Ratio (TSR) Control.

We know that the power produced at any instance depends on torque (T) and angular velocity (ω). Therefore, using Equation (21), it is easily perceivable that to optimal power (P_O) can be obtained given in Equation (23) similarly by multiplying the result with the term of angular velocity, applying minor modifications to it. Since here the rotor speed is monitored, the angular velocity of the generator has to be replaced by the angular velocity of the rotor.

$$P_O = K_{op} {\omega_T}^3 \tag{23}$$

Here, K_{op} is a design constant, based on physical parameters of the turbine. Hence, Equation (23) gives the output as the reference power. Figure 8 illustrates this control through a block diagram [53].

Figure 8. Power Signal Feedback Control Block Diagram.

Over recent years, various pieces of research have been conducted to further utilize the potential of control strategies. As a result, improvisation of already existing methods to increase effectiveness has been brought up.

Zhang et al. presented a fuzzy logic controller for the pitch angle of the turbine [56]. Traditional pitch control logics employ the use of a PI controller, as discussed in level 4 control. The implication of such controller is that the user or operator needs to have knowledge of mathematical model of the same. However, with the application of a fuzzy logic, the wind speed can be used to compensate for the non-linear sensitivity. The use of fuzzy logic controllers is apparent where the system dynamics contain multiple non-linearities like inconsistent wind speeds. The presented study incorporates triangular symmetrical membership functions for input and output parameters, as well as a mathematical model of the turbine created using MATLAB. The triangular symmetrical functions provide increased sensitivity for variable values that are approaching zero. The width of the functions can be adjusted with respect to each parameter. The outcome of the alternate fuzzy logic helped in reducing the ripples in torque as seen in conventional PID controllers, signifying a smoother torque curve.

Another MPPT control strategy was proposed by Dalala et al., in 2013, and suggested maximizing the energy captured while operating at wind speeds below the rated wind speed. This study is applicable for smaller scale WECS [57]. This strategy of control involves multiple stall controls for high wind speed regions. A constant speed stall controller allows the rotational speed of the generator to be kept limited to the rated speed, while a constant power stall controller helps to keep the generated power within permissible limits. This control strategy works on two inputs, namely, the electromechanical torque and the wind speed. To compensate for the instability of stall regions, the proposed control employs a stabilizing control loop, and a mode of transfer control strategy is innovated to stabilize the transition between various operational modes.

6. Present Trends in the Field of WECS and Scopes of Improvement

6.1. On-Going Trends

A sharp incline in the installation of wind energy firms have been observed within recent decades throughout the world, rendering a shift of dependence on clean energy [58]. The present trend of global wind energy production has been represented in Figure 9 [59,60]. In India itself, the total wind energy capacity was found to be 21,263 MW until May 2014. With the harnessing of wind energy on such large scale, recent times have attracted research and development on every aspect of WECS.

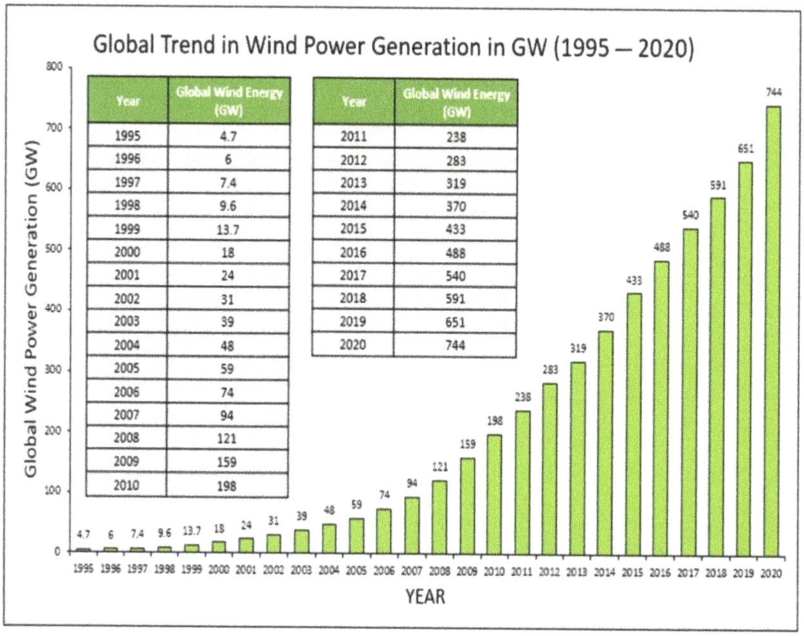

Figure 9. Trend in Wind Power Generation in Giga Watt (GW) [59,60].

Modern wind energy conversion systems rely almost totally on power produced by horizontal axis wind turbines. Such wind turbines are common scenarios in large commercially- and grid-connected wind farms, while the development of vertical axis counterparts have been impeded early, as they were in the later decades of 20th century. The concentration of these studies took place only in the US and UK, at the US Department of Energy Sandia Lab and Reading University, respectively [58]. However, further study on the topic ceased once it was established that VAWTs were incapable of producing power as efficiently as HAWTs could for large scale power production. The general concept of HAWT and VAWT gives way to few disadvantages for the latter. VAWTs are inefficient, as the blades move approximately half of the rotation along the direction of wind, and the other half against the same wind. This is not much of a problem for HAWTs as the part of the rotor going against the direction of wind is far below the point that traverses along the direction of the wind. Since wind at the top is stronger, due to characteristics of wind with altitude, the force driving the turbine is far higher than the force that opposes it. Another main advantage of HAWTs over VAWTs is that they can be installed at any height, however in contrast, the VAWTs must be installed close to the ground. Therefore, an installation that is not on a mast loses a great deal of efficiency in case of VAWTs due to the fact that natural wind is blown more strongly and evenly at greater heights [61].

Contrary to old research, recent developments have suggested the practicality of vertical axis turbines in the large-scale generation of power in wind firms, as they have better aerodynamic efficiency than the traditional commercial HAWTs [62]. Vertical axis wind turbines are brought back into focus for evaluation of its aerodynamic properties as well as their performances regarding flow separation. Attempts to nullify the adverse effects of energy production using these set ups are practiced [63]. Furthermore, to test the endurance and reliability of a VAWT and to meet the energy requirements, recently a VAWT has been designed according to the harsh conditions of the Amundsen–Scott South Pole Station [64], which is a scientific research station in southernmost place on the globe. Being at the extreme end of South Pole, the turbine is subjected to rather harsh weather conditions.

As stated in the previous section, the majority of research presently being conducted is to test the application of various control strategies and methodologies with different set ups. With the advancement in the past decade of software and computer technologies, the research regarding control strategies highly incorporates simulation studies. The simulations carried out for confirming the effectiveness involve a mathematical model of the turbine itself. Usage of Maximum Power Point Tracking controller for generating control strategies are focused primarily on recent studies, which is evident due to the simplicity and potential of the system [65]. As DFIG has a low inertia response, they are used in most wind turbines. The response of these DFIGs can be modified for good with the inclusion of a feedback parameter, where this would be used as a function of the rate of change of the frequency of the grid with respect to the electromagnetic torque [66]. As a response to the adaptation of DFIG for commercial purposes, it is imperative to integrate the potential of artificial intelligence (AI), for actively configuring the operational parameters. AI is being used in all aspects of society at present. Studies have shown interesting results, comparing the robustness and response time of an artificial neural network controller to a classical PI controller [67]. In every case, it has been found that the implementation of a neural network to predict the wind flow patterns have rendered more effective control. The response time could also be significantly reduced, with a higher robust operation.

Advancements in energy storage technologies are at a peak as well [64]. Even though for modern wind turbines, being directly connected to the grid does not directly require a power storage module. However, there are components that incorporate the need of batteries, such as in ancillary services that act as a backup power source for running the control systems, besides making the output to the grid far more predictable. While traditionally many devices and methodologies have been formulated to flatten the generated power curve so as to get rid of spikes in it, for the waveform received in the grid side, efficient storage technologies can also help to provide a smooth output power curve [68]. Storage technologies being investigated in recent years to find their application in wind energy systems include the following, although this list is not exhaustive [56]:

- battery;
- flow battery (FB);
- fuel cell (FC);
- flywheel (FW);
- superconducting magnetic energy storage (SMES);
- supercapacitor (SC);
- compressed air energy storage (CAES);
- pumped hydroelectric energy storage (PHES);
- water storage.

In contrast to flywheels that store energy by virtue of mechanics [69], FC and battery can store energy in the form of chemical energy, making them more practical in the intended applications. Promising technologies of battery for the use in grid-energy systems include advanced lead–acid (Pb–acid), nickel–cadmium (Ni–Cd), lithium-ion (Li+), sodium–sulphur (Na–S), nickel metal hydride (NiH) and flow batteries [63].

a. Advanced lead-acid battery: Regular lead–acid batteries have been a staple for a variety of applications in every corner of industry, as well as in an automotive and domestic context. Though, most prominent and specific applications include frequency regulation, bulk storage of energy in case of integration of variable renewable energy sources and distributed energy storage systems. As a result of versatile and vast usage of the Pb–acid batteries, these have been an interesting focus for further enhancements. A conventional Pb–acid battery comprises lead–dioxide (PbO_2) as cathode and lead (Pb) as anode. During the discharge process, chemical reactions in the cell lead to the synthesis of lead sulfate, whereas, during charging, this reaction reverses and Pb gets deposited back onto the electrodes. Recent developments have demonstrated that introducing carbon into the cell can help with the prevention of detrimental crystallization during charging and discharging cycles [56]. The implication of this is an increase in the life-span of the batteries, besides an increase in life-cycle between consecutive charging.
b. Sodium-sulfur battery: The core of this battery technology houses molten sodium in an ionic state, which is lined with alumina (or aluminum oxide) on all sides. This is placed inside a sodium sulfate surrounding. The advantage of sodium sulphur battery is a high energy density, making it suitable for places of application that demand short yet potent bursts of energy. These batteries are used in rather functional contexts in energy distribution grid applications, wind power integration, in ancillary services as UPS batteries and much more.
c. Lithium ion battery: As the name points out, these batteries incorporate positively charged lithium ions (as metal), that flow through electrolyte and promotes flow of electrons across the connected outer circuit. From time to time, lithium ion batteries have found a place in almost all electronic applications, alongside medical uses, owing to the far superior energy density values. At present, lithium ion batteries are being tried and tested for application in automotive industries, prompting its upcoming uses in fields of renewable energy industry [56]. Current developments are focused on improving the charging and discharging cycle efficiency of these batteries.
d. Flow batteries: A flow battery technology stores energy by the virtue of chemical energy in the form of two electrolyte reservoirs inside the body of the battery. The chemical energy is converted into electrical energy when one liquid is displaced slowly and steadily from one tank to another. The displacement of the liquid when reversed acts to charge the batteries. Since they are the function of the fluid motion, they are termed as flow batteries. The energy storing capacity of these batteries is solely governed by the size of the tanks and the energy density is dependent on the

volume flow rate from anode to cathode. The primary advantage, for which they are considered as a potential candidate for evolving battery technologies, comes from the fact that these batteries can have very high-power ratings, and the electrolytes can be stored and replaced easily, more conveniently than typical cell electrolytes, coupled with adjustable and replaceable flow properties according to the application.

The water storage is an unused generated energy, which is trying to be converted into potential energy from the water bring pumped from a lower container to a higher one. In the near future, its performance might be compared with chemical batteries [70].

Adaption of newer technologies has always been a lengthy process. Usually, a new technology is expensive to begin with, making them undesirable at this early stage. However, with lots of scope for improvement, those who take up the idea and make it the focus of further studies and increase understanding make them cheaper and more acceptable, besides making them superior with regard to existing technologies [71]. In this context, the novel nanomaterials would play a vital role in the near future of these renewable systems [72,73]. For example, superhydrophobic coating would enhance the shelf-life of the turbine, as well as various components of the WECS [74]. The emission properties of nanolayers could be used to investigate the erosion scars of the turbines [75,76]. Rapid adaptation of new technology in the market, coupled with the replacement of older ones, results in the effective penetration of a new product of about 15% [56]. A similar phase is being observed by wind turbine technologies.

6.2. Future Prospects

Since wind energy is a relatively newer source of energy, most of the resources that capture the energies have focused on the development of turbines and tower designs, as well as methods to increase the efficiency of existing systems. However, possibilities for the evolution of technologies are endless. However, a future emerging technology in this field of study would consider the following criteria to be accepted [77]:

a. Technologies based on supply and conversion of wind energy;
b. A technical overhaul, with new fundamentals, which cannot be a result of incremental research on prevalent technologies; and
c. A technology which is in its early developmental stages.

While some of the technologies enlisted below have been used in practice, they may not have necessarily been used in wind energy systems and thus have ample space for improvements to be made usable. Even if not a completely radical innovation, large scale adaptation of such technologies might be considered as an emerging technology.

6.2.1. Airborne Wind Energy (AWE)

AWE systems are a collective group of devices that harness energy in a similar form as a regular WECS set-up, instead having the external features of an unmanned aircraft or an autonomous kite. An airborne wind turbine is schematically shown in Figure 10 [78]. The system is suspended in air and attached to the ground by means of one or more tethers to keep it in place. This system has some advantages over modern WECS technologies, including a huge reduction in the material needed for construction, since it lacks the requirement of a tower. There have been many concept designs developed for this aspect of power generation, however none of them have been finalized.

The two broad categories of the AWE system involve ground-gen and fly-gen systems. The ground-gen concept aims at converting the mechanical power into electrical power at the ground level, while fly-gen systems do so in the airborne unit itself. The fly-gen systems are further classified into crosswind and non-crosswind, out of which the crosswind turbines have been observed to produce more power [78]. Besides the advantage of lower cost owing to the lesser materials for construction, the airborne wind turbine can be raised high for interacting with high-altitude winds found above 200 m [79]. Simultaneously, the altitudes can be altered to protect it from winds above rated speeds.

However, there are many technical challenges associated that are yet to be overcome for making this usable, such as reliability and durability of flexible components and erosion by wind and water. Recent trends in the research of this system consist of drone concepts, aerodynamic modeling, electronics and sensor technologies for autonomous control of the tethered devices, and controls for autonomous take-off and landing.

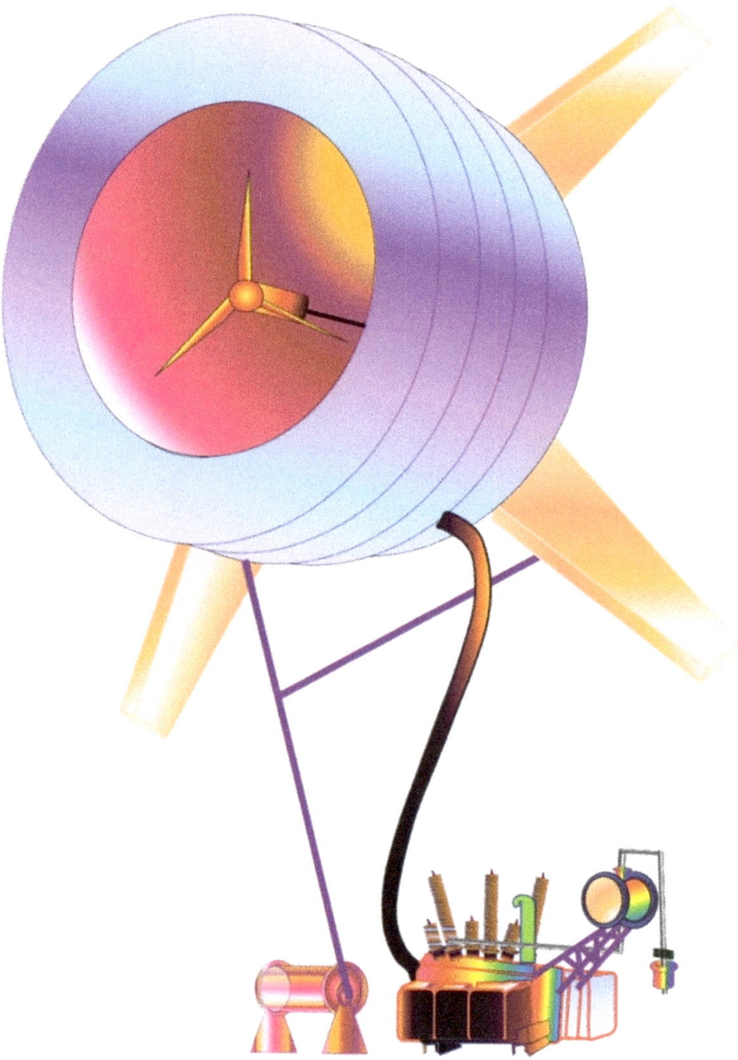

Figure 10. Airborne Wind Turbine [78].

6.2.2. Offshore Floating Wind Concepts

This concept differs from traditional offshore concepts by the method of foundation. While the latter employs a fixed foundation to the floor of the ocean, this innovative design is a free-floating structure related to the semi-submersible tension leg or the spar platforms. The floating is controlled by various anchoring systems. Since it is a diversion from the existing fixed off-shore designs, there high potential for the exploration of newer techniques to integrate the tower to the floating base. Moreover, the impact of changing turbine parameters including TSR, blade designs and number of blades on the floating platform can be investigated [80].

Recent progress in this topic has led to evolution of three main types of floating off-shore foundations including spar-buoy, tension leg platform, and spar-submersible, which are depicted in Figure 11A–C, respectively, and is briefly explained below [81]:

Spar-buoy: Stability in this set-up is provided by a large cylindrical buoy filled with ballast under the tower structure. This helps in lowering the center of gravity (CG) in the system below the level of the center of buoyancy and increases the stability. Apart from the heavy buoy under the water, the hollow structure above helps to keep the center of buoyancy high.

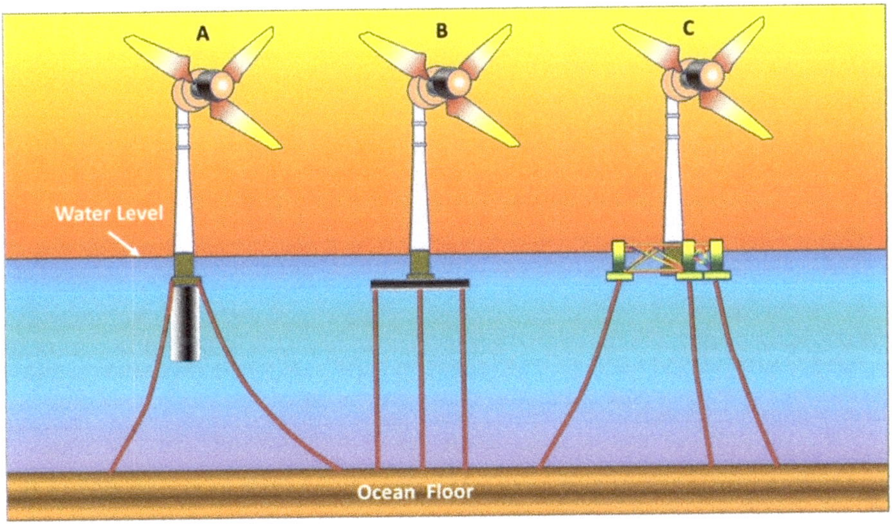

Figure 11. (**A**) Spar-buoy; (**B**) Tension Leg Platform; (**C**) Spar-Submersible [80,81].

Tension Leg Platform: This design incorporates the buoyant component to be semi-submerged into the ocean surface, which is then attached to the ocean floor by means of tension mooring lines, providing stability to the tower above it. The buoyant structure is maintained; submerged by balancing out the upward buoyant force. it experiences by the mooring lines' tension.

Spar-Submersible: Also called the semi-submersible design, this concept incorporates multiple columns that are linked by bracings or submersible pontoons. The columns provide hydrostatic stability to the structure, whereas the pontoons provide additional buoyancy to it. This system also uses catenary mooring lines to keep it in place [78].

6.2.3. Tip-Rotor Wind Turbine (TRWT)

This concept of wind turbine system integrates high speed, low torque rotors at the tip of the main blades. Therefore, the generator can be coupled directly without the necessity of torque reduction gearbox arrangements for these rotors. The efficiency of these rotors to extract wind energy can go beyond the Betz limit, since it is actually the thrust of these tip-rotors that provides reaction torque to the main rotor of the turbine [82]. The rotation of the bigger main rotor forces the smaller rotors to move and thus harness power. Tests have been conducted in this field by companies making commercial turbines. Siemens had demonstrated a concept of direct drive for e-propellers for an aircraft that could reach performance levels of about 5 kW/kg [83]. Moreover, this signifies that a 10 MW turbine using two blades incorporating this concept, can operate using two generators of 1000 kg each housed at the tip of the blades. This can massively reduce the weight of commercial MW WECS, saving capital costs a huge volume.

6.2.4. Multi-Rotor Wind Turbine (MRWT)

Replacing the large size of a rotor with multiple smaller rotors is helpful to reduce the overall loads on the structure of the turbine. This concept made it possible to install a large power system of about 20-MW at a single site. A company like, Vestas, has adapted this concept to develop a 900 kW four rotor design, illustrated in Figure 12. The most prominent advantage of this technology is the process of standardization as it allows the production process of these small-scale rotors to be industrialized [83]. This process could help is saving production costs in comparison to present large scale rotors, since these large-scale rotors need to be customized for specific application, thus reducing probability of standardization. Studies have enlightened the advantages in maintenance by having smaller components.

The modelling of multi-rotor system needs more development owing to its challenging aspect, as well as aerodynamic studies regarding effects of wake produced by each rotor on adjacent ones need to be looked further, since the design tools for simulating aerodynamic behavior required to visualize a multi-rotor system and turbulent winds are in their elementary stages. The overall design of the setup is also worth looking into for improving the reliability, which is curtailed by the increased static loads and dynamic vibration of increased number of components [84].

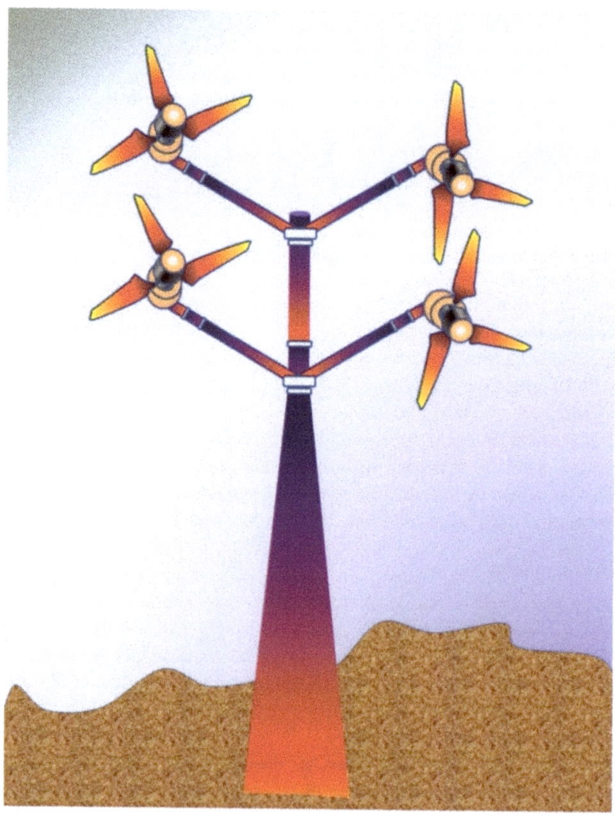

Figure 12. Multi-rotor Wind Turbine [83].

6.2.5. Diffuser Augmented Wind Turbine (DAWT)

Diffuser Augmented Wind Turbines, chiefly abbreviated as DAWT is a modified design of normal horizontal axis wind turbines, with the addition of a diffuser structure around the rotor. The diffuser is encased around with a diffuser that resembles the shape of a funnel, used to concentrate the approaching stream of wind. Modification can be introduced to this structure by adding a rim around the exit side of the diffuser and an inlet shroud at the entrance side (upwind). The diffuser has been proven to increase the generated power compared to traditional HAWTs [85]. A schematic diffuser augmented wind turbine is shown in Figure 13 [86].

Figure 13. A schematic diagram diffuser augmented wind turbine [86].

The concept of DAWTs is being developed in Japan at a semi-commercial level. The rating of these is in a range of ten kW at present [83], with the most efficient design developed by Kyushu University, who have developed a DAWT system capable of producing 100 kW. This is however the outcome of extensive study and testing of various configurations of DAWTs [86]. This design is an integration of a multi-rotor turbine and a DAWT, effectively making a multi-rotor diffuser augmented turbine. Multi-rotor DAWT systems have been demonstrated to have performance advantage over multi-rotor turbines. Optimization in space between each diffuser can have significant advantage in the overall design extracted power, since the geometric symmetrical structure may not be really advantageous in the intended use, due to flow parameters of free stream.

With a more complex design than its counterpart, it may not be a cheap alternative. However, if it is used properly, it would be suitable in certain situations. At present, diffuser augmented turbines face the biggest challenges of structural loading. Though this design has sound potential, they are best suited for small scale applications.

7. Material Aspects

In wind energy conversion systems, both mechanical and electrical components play a vital role. Therefore, the materials used for mechanical and electrical components in the WECS are also extremely crucial. Specifically, the materials used for mechanical parts help to get 'work' of the system using a mass to rotate the turbine blade and the materials used for electrical parts help to transmit the mass to a desired location. Typically, a wind turbine is constructed of nearly 25,000 elements, which are grouped into several major components including, tower, nacelle and rotor [87]. The rotor, which contains blades, a hub and a blade pitch system, is attached to a nacelle. The nacelle is connected to the tower. The nacelle comprises numerous electrical and mechanical components, such as main a shaft, a gearbox, generator and control systems. As the turbine blades rotate, it leads to drive an electrical generator through a gearbox. The gear box contains special alloys to adapt wide range of wind speeds [88]. The gear box is placed on top of a tower, which is made of large tubular steel sections. The steel sections are mainly attached to an anchor component and constructed on a foundation [89]. In addition, the wind turbine plant also connected some conventional ground-mounted components including the plant transformer, switchgears, and site cables, which connects between the turbines, to the transformer and to the grid.

In a rotor subassembly, the rotating body comprises rotor blades, rotor bearings, nose cone, hubs, and the pitch drive system. Therefore, the major mechanical components of the WECS, for which the selection of material is very important, are rotor, main shaft, gearbox, mechanical brakes, nacelle, pitch and yaw drives, and wind measuring equipment. On the other hand, the major electrical components of the WECS, for which the selection of material is very important, are generator, power converter, step-up transformer, and so on. The electrical subsystems convert the mechanical energy into electrical energy from the wind. The most useful components employed for the wind turbine energy supply chains from the necessary raw materials, processed materials, components and up to the final products are depicted in Figure 14 [90].

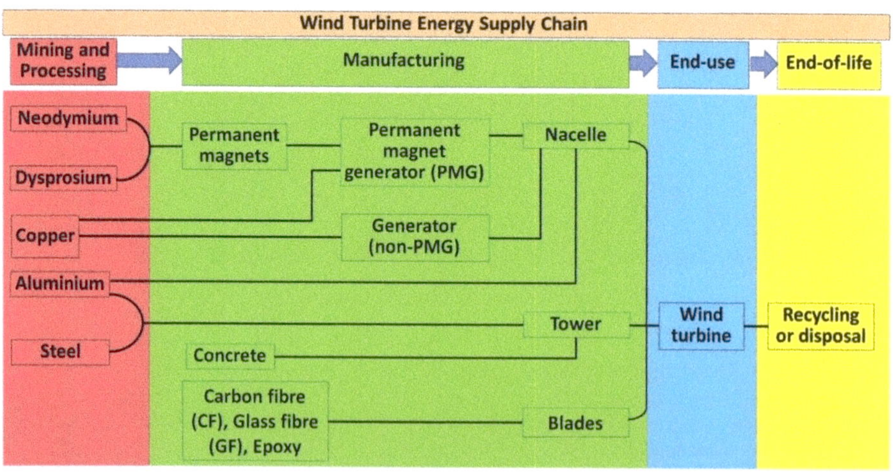

Figure 14. The most useful components used for the wind turbine energy supply chains from the necessary raw materials, processed materials and components, to the final products.

In the WECS, the wind turbine with efficient wing construction is extremely important to the material scientists. The various components that are used in a wind turbine for renewable energy generation are supportive materials of the turbine, struts, shaft, bearings, chain, sprockets, gear, turbine blades, motor components, tower and so on. In this aspect, NASA invented many turbines with various materials and different technologies, which were mainly aimed at improving energy conversion. The most recent research on wind turbine focused on composite blade materials, steel tube towers, variable-speed generators, and dynamic components including aerodynamics of the blades in horizontal axis turbines, pitch and span control mechanisms. Typically, the weight sharing of an onshore wind turbine plant consists of 75% foundations, 23% turbine, and 2% in cite cables. The turbine part comprises the 59 wt% tower, 22 wt% nacelle, and 19 wt% rotor. The mass distribution of each important elements of wind turbine plant is illustrated in Table 2. Many components, such as tower, nacelle, rotor and foundation are made of iron particularly, mild steel and stainless steel. In steels, a wide range of minor and base metals such as nickel (Ni), chromium (Cr), molybdenum (Mo), and manganese (Mn) are used.

Table 2. Mass distribution of each important element of wind turbine plant.

Materials	Mass (%)	Components
Concrete	50–60	Foundation
Steel	25–35	Foundation, tower, nacelle, generator, gear box
Composite materials	2–3	Turbine blades
Electronic components	<1	Generator
Copper (Cu)	<1	Foundation, tower, nacelle, generator, gear box
Aluminium (Al)	<1	Tower, nacelle
Polyvinyl chloride (PVC)	<1	Foundation, Turbine blades
Lubricant and other liquids	<1	Gearbox, generator
Self-cleaning coatings	<1	Turbine blades, tower, gear box

For several years, scientists have been trying to find suitable materials for rotor blades. Earlier the rotor blades were made from heavier metals. These high density or higher mass metals are resistant to environmental conditions such as high turbulent wind, however are rendered not helpful as a result of their large weight. In addition, the different mass concentrations away from the center create larger rotational moment of inertia. This causes a reduction in the acceleration of the rotor. This difficulty led the focus of the material towards composites, which have high rigidity without increasing the mass. Therefore, the advantages, disadvantages, scopes and challenges of the conventional as well as natural fiber reinforced composite materials used for the wind turbine blades are depicted in Figure 15.

In addition, different advanced smart materials such as, piezoelectric materials are trying to use for "wind tree" in order to produce energy or power from the wind flow. The soft flexible materials are used as a leaf element to generate maximum power. The trunk parts of the wind tree-like structure are made of hard materials such as steel. The steel trunks stem into tinier branches, which are attached with many leaf-shaped wind turbines. These wind trees have recently been exploited to all types of wind, from gentle breezes to powerful gusts, in both urban and rural areas [91].

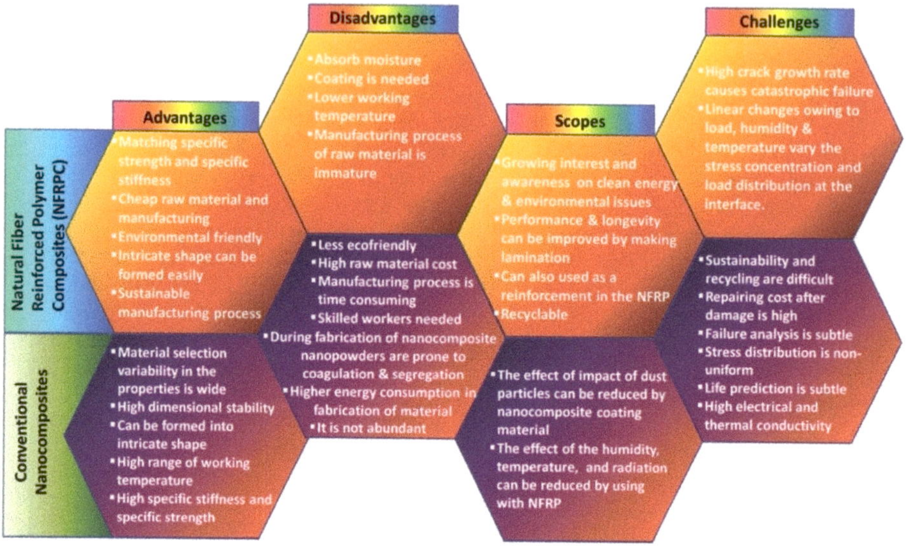

Figure 15. The advantages, disadvantages, scopes and challenges of the conventional as well as natural fiber reinforced composite materials used for the wind turbine blades.

8. Conclusions

A long way has been traversed, with an array of technologies implemented in building the modern-day wind energy conversion systems. However, it has long list of prospects in the upcoming future. Shifting the focus away from the carbon-based fossil fuel reserves has opened up a multitude of opportunities waiting to be explored. This review paper discussed in detail the mechanical and electrical components of conventional WECSs, along with the various alterations in the designs, which have helped in categorization of the systems. This review paid much attention to the ecological applications of WECS and wind power plants. It is also to be informed that the hydropower plant could be another alternative sustainable power source since wind flow is a less stable power source. For wind power generation, the water consumption is lowest. It has the lowest relative greenhouse gas emission. Thus, it has the most favorable social impacts and is considered as one of the most sustainable renewable energy sources, followed by hydropower, photovoltaic, and geothermal.

Most commercial applications have adopted HAWT designs for WECS over VAWTs due to the former's compatibility with large scale power production. Different control techniques and strategies were discussed. A majority of wind turbines in the present context operate on doubly fed induction generators, the control strategies of which have been tabulated expressing the salient features of each design. In the context of the present scenario, the evolving technologies for harnessing power from wind have been highlighted, and the future emerging technologies have also been addressed. The reliability of wind energy has increased exponentially over the last few decades. The expansion of wind energy along with other renewable energy sources will continue to grow in the near future as well.

We have found that materials for the blade and generator are the most critical elements of wind turbines. Since many components of wind power plants use synthetic polymer matrix composite materials, the recycling of composite is a cross-sector challenge. This huge challenge is not only a problem for the wind power plants, but also to its supply chain industries. Therefore, all the sectors connected to synthetic polymer matrix composite materials must look into the matter together in order to develop cost-effective solutions and value chains for the combined volume of composite waste.

Author Contributions: Conceptualization, methodology, and resources, S.P.; writing—original draft preparation, A.C., R.D., M.P.K. and S.P.; writing—review and editing, S.P. and J.P.D.; supervision, S.P. All authors have read and agreed to the published version of the manuscript.

Funding: This research received no external funding.

Institutional Review Board Statement: Not applicable.

Informed Consent Statement: Not applicable.

Data Availability Statement: Not applicable.

Conflicts of Interest: Authors do not have any conflict of interest.

References

1. Speirs, J.; McGlade, C.; Slade, R. Uncertainty in the availability of natural resources: Fossil fuels, critical metals and biomass. *Energy Policy* **2015**, *87*, 654–664. [CrossRef]
2. Sukhatme, S.P.; Nayak, J. *Solar Energy*; McGraw-Hill Education: New York, NY, USA, 2017.
3. Herbert, G.J.; Iniyan, S.; Sreevalsan, E.; Rajapandian, S. A review of wind energy technologies. *Renew. Sustain. Energy Rev.* **2007**, *11*, 1117–1145. [CrossRef]
4. Balat, M.; Balat, H. Biogas as a renewable energy source—A review. *Energy Sources Part A* **2009**, *31*, 1280–1293. [CrossRef]
5. Melikoglu, M. Current status and future of ocean energy sources: A global review. *Ocean Eng.* **2018**, *148*, 563–573. [CrossRef]
6. Blunden, L.; Bahaj, A. Tidal energy resource assessment for tidal stream generators. *Proc. Inst. Mech. Eng. Part A J. Power Energy* **2007**, *221*, 137–146. [CrossRef]
7. Delyannis, E. Historic background of desalination and renewable energies. *Sol. Energy* **2003**, *75*, 357–366. [CrossRef]
8. Redlinger, R.; Andersen, P.; Morthorst, P. *Wind Energy in the 21st Century: Economics, Policy, Technology and the Changing Electricity Industry*; Springer: New York, NY, USA, 2016.
9. Ajao, K.R.; Mahamood, M.R. Wind energy conversion system: The past, the present and the prospect. *J. Am. Sci.* **2009**, *5*, 17–22.
10. Ellabban, O.; Abu-Rub, H.; Blaabjerg, F. Renewable energy resources: Current status, future prospects and their enabling technology. *Renew. Sustain. Energy Rev.* **2014**, *39*, 748–764. [CrossRef]
11. Kumar, R.; Ganesh, A.; Kumar, V. Raising concerns on high PV penetration and ancillary services: A review. In Proceedings of the International Conference on Intelligent Computing and Smart Communication, Ostrava, Czech Republic, 21–22 March 2019; Springer: Singapore, 2020; pp. 1347–1364.
12. Pedraza, J.M. The current situation and perspectives on the use of solar energy for electricity generation. In *Electrical Energy Generation in Europe*; Springer: Cham, Switzerland, 2015; pp. 169–219. [CrossRef]
13. Ahmadi, G. Some preliminary results on the performance of a small vertical-axis cylindrical wind turbine. *Wind Eng.* **1978**, *2*, 65–74.
14. Hepbasli, A.; Ozgener, O. A review on the development of wind energy in Turkey. *Renew. Sustain. Energy Rev.* **2004**, *8*, 257–276. [CrossRef]
15. Jaber, J.O.; Badran, O.O.; Abu-Shikhah, N. Sustainable energy and environmental impact: Role of renewables as clean and secure source of energy for the 21st century in Jordan. *Clean Technol. Environ. Policy* **2004**, *6*, 174–186. [CrossRef]
16. Howey, D.A.; Bansal, A.; Holmes, A.S. Design and performance of a centimetre-scale shrouded wind turbine for energy harvesting. *Smart Mater. Struct.* **2011**, *20*, 085021. [CrossRef]
17. Shah, L.; Cruden, A.; Williams, B.W. A variable speed magnetic gear box using contra-rotating input shafts. *IEEE Trans. Magn.* **2010**, *47*, 431–438. [CrossRef]
18. Price, T.J. James Blyth—Britain's first modern wind power pioneer. *Wing. Eng.* **2005**, *29*, 191–200. [CrossRef]
19. Horizontal Axis Wind Turbine. Available online: https://www.secret-bases.co.uk/wiki/Horizontal_axis_wind_turbine (accessed on 1 December 2021).
20. Brunt, A.; Spooner, D. The development of wind power in Denmark and the UK. *Energy Environ.* **1998**, *9*, 279–296. [CrossRef]
21. Nelson, V.; Starcher, K. *Wind Energy: Renewable Energy and the Environment*; CRC Press: Boca Raton, FL, USA, 2018. [CrossRef]
22. Hand, B.; Cashman, A. A review on the historical development of the lift-type vertical axis wind turbine: From onshore to offshore floating application. *Sustain. Energy Technol. Assess.* **2020**, *38*, 100646. [CrossRef]
23. Sharifian, M. The Paths of Clean Technology: From Innovation to Commercialization. Ph.D. Thesis, University of Alberta, Edmonton, AB, Canada, 2015.
24. Von Gersdorff, K. Transfer of German aeronautical knowledge after 1945. In *Aeronautical Research in Germany*; Springer: Berlin/Heidelberg, Germany, 2004; pp. 325–344. [CrossRef]
25. Wang, W.; Caro, S.; Bennis, F.; Mejia, O.R.S. A simplified morphing blade for horizontal axis wind turbines. *J. Sol. Energy Eng.* **2014**, *136*, 011018. [CrossRef]
26. Nielsen, K.H. Technological trajectories in the making: Two case studies from the contemporary history of wind power. *Centaurus* **2010**, *52*, 175–205. [CrossRef]

27. Jaen-Sola, P.; McDonald, A.S.; Oterkus, E. Lightweight design of direct-drive wind turbine electrical generators: A comparison between steel and composite material structures. *Ocean Eng.* **2019**, *181*, 330–341. [CrossRef]
28. Yaramasu, V.; Wu, B. *Model Predictive Control of Wind Energy Conversion Systems*; John Wiley & Sons: Hoboken, NJ, USA, 2016.
29. Shkara, Y.; Cardaun, M.; Schelenz, R.; Jacobs, G. Aeroelastic response of a multi-megawatt upwind horizontal axis wind turbine (HAWT) based on fluid–structure interaction simulation. *Wind Energy Sci.* **2020**, *5*, 141–154. [CrossRef]
30. Didane, D.H.; Maksud, S.M.; Zulkafli, M.F.; Rosly, N.; Shamsudin, S.S.; Khalid, A. Performance investigation of a small Savonius-Darrius counter-rotating vertical-axis wind turbine. *Int. J. Energy Res.* **2020**, *44*, 9309–9316. [CrossRef]
31. Yao, J.; Liao, Y. Analysis on the operations of an AC excited wind energy conversion system with crowbar protection. *Autom. Electr. Power Syst.* **2007**, *31*, 79–83.
32. Buehring, I.; Freris, L. Control policies for wind-energy conversion systems. *IEE Proc. C Gener. Transm. Distrib.* **1981**, *128*, 253–261. [CrossRef]
33. Chang, L. Wind energy conversion systems. *IEEE Can. Rev.* **2002**, 12–16.
34. Schubel, P.J.; Crossley, R.J. Wind turbine blade design. *Energies* **2012**, *5*, 3425–3449. [CrossRef]
35. Nejad, A.R.; Jiang, Z.; Gao, Z.; Moan, T. Drivetrain load effects in a 5-MW bottom-fixed wind turbine under blade-pitch fault condition and emergency shutdown. *J. Phys. Conf. Ser.* **2016**, *753*, 112011. [CrossRef]
36. Salameh, J.P.; Cauet, S.; Etien, E.; Sakout, A.; Rambault, L. Gearbox condition monitoring in wind turbines: A review. *Mech. Syst. Signal Process.* **2018**, *111*, 251–264. [CrossRef]
37. Manolesos, M.; Chaviaropoulos, P. Wind tunnel study of a generic wind turbine nacelle model. In *Turbo Expo: Power for Land, Sea, and Air, Proceedings of the Turbomachinery Technical Conference and Exposition, Charlotte, NC, USA, 26–30 June 2017*; American Society of Mechanical Engineers: New York, NY, USA, 2017; p. V009T49A003.
38. Blaabjerg, F.; Chen, Z.; Teodorescu, R.; Iov, F. Power electronics in wind turbine systems. In Proceedings of the 2006 CES/IEEE 5th International Power Electronics and Motion Control Conference, Shanghai, China, 14–16 August 2006; pp. 1–11.
39. Abbasi, M.; Lam, J. A step-up transformerless, ZV–ZCS high-gain DC/DC converter with output voltage regulation using modular step-up resonant cells for DC grid in wind systems. *IEEE J. Emerg. Sel. Top. Power Electron.* **2017**, *5*, 1102–1121. [CrossRef]
40. Jose, G.; Chacko, R. A review on wind turbine transformers. In Proceedings of the 2014 Annual International Conference on Emerging Research Areas: Magnetics, Machines and Drives (AICERA/iCMMD), Kottayam, India, 24–26 July 2014; pp. 1–7.
41. Islam, M.R.; Guo, Y.; Zhu, J. Multilevel converters for step-up-transformer-less direct integration of renewable generation units with medium voltage smart microgrids. In *Large Scale Renewable Power Generation*; Springer: Singapore, 2014; pp. 127–149. [CrossRef]
42. Ragheb, M.; Ragheb, A.M. Wind Turbines Theory—The Betz Equation and Optimal Rotor Tip Speed Ratio. In *Fundamental and Advanced Topics in Wind Power*; IntechOpen: London, UK, 2011. [CrossRef]
43. Huleihil, M.; Mazor, G. Wind turbine power: The betz limit and beyond. In *Advances in Wind Power*; IntechOpen: London, UK, 2012. [CrossRef]
44. Ragheb, M.; Ragheb, A.M. Wind turbines theory-the betz equation and optimal rotor tip speed ratio. *Fundam. Adv. Top. Wind Power* **2011**, *1*, 19–38.
45. Whitaker, S. Newton's laws, Euler's laws and the speed of light. *Chem. Eng. Educ.* **2009**, *43*, 96–103.
46. Ranjbar, M.H.; Nasrazadani, S.A.; Kia, H.Z.; Gharali, K. Reaching the betz limit experimentally and numerically. *Energy Equip. Syst.* **2019**, *7*, 271–278.
47. Burton, T.; Sharpe, D.; Jenkins, N. *Handbook of Wind Energy*; John Wiley & Sons: Hoboken, NJ, USA, 2001.
48. Nguyen, H.M.; Naidu, D.S. Advanced control strategies for wind energy systems: An overview. In Proceedings of the 2011 IEEE/PES Power Systems Conference and Exposition, Phoenix, AZ, USA, 20–23 March 2011; pp. 1–8. [CrossRef]
49. Teodorescu, R.; Liserre, M.; Rodríguez, P. *Grid Converters for Photovoltaic and Wind Power Systems*; John Wiley & Sons: Hoboken, NJ, USA, 2011.
50. Song, D.; Yang, J.; Su, M.; Liu, A.; Liu, Y.; Joo, Y.H. A Comparison study between two MPPT control methods for a large variable-speed wind turbine under different wind speed characteristics. *Energies* **2017**, *10*, 613. [CrossRef]
51. Ananth, D.; Kumar, G.V.N. Tip speed ratio based MPPT algorithm and improved field oriented control for extracting optimal real power and independent reactive power control for grid connected doubly fed induction generator. *Int. J. Electr. Comput. Eng.* **2016**, *6*, 1319.
52. Blaabjerg, F.; Iov, F.; Chen, Z.; Ma, K. Power electronics and controls for wind turbine systems. In Proceedings of the 2010 IEEE International Energy Conference, Manama, Bahrein, 18–22 December 2010; pp. 333–344.
53. Thongam, J.S.; Ouhrouche, M. MPPT control methods in wind energy conversion systems. *Fundam. Adv. Top. Wind Power* **2011**, *15*, 339–360.
54. Barbosa de Alencar, D.; De Mattos Affonso, C.; Limão de Oliveira, R.C.; Moya Rodríguez, J.L.; Leite, J.C.; Reston Filho, J.C. Different models for forecasting wind power generation: Case study. *Energies* **2017**, *10*, 1976. [CrossRef]
55. Phan, T.; Nguyen, V.; Hossain, M.; To, A.; Tran, H.; Phan, T. Transient responses of the doubly-fed induction generator wind turbine under grid fault conditions. In Proceedings of the 2016 International Conference on Advanced Computing and Applications (ACOMP), Can Tho, Vietnam, 23–25 November 2016; pp. 97–104.
56. Islam, M.R.; Mekhilef, S.; Saidur, R. Progress and recent trends of wind energy technology. *Renew. Sustain. Energy Rev.* **2013**, *21*, 456–468. [CrossRef]

57. Dalala, Z.M.; Zahid, Z.U.; Lai, J.-S. New overall control strategy for small-scale WECS in MPPT and stall regions with mode transfer control. *IEEE Trans. Energy Convers.* **2013**, *28*, 1082–1092. [CrossRef]
58. Price, T.J. UK Large-scale wind power programme from 1970 to 1990: The Carmarthen Bay experiments and the musgrove vertical-axis turbines. *Wind Eng.* **2006**, *30*, 225–242. [CrossRef]
59. Biswal, G.; Shukla, S. Recent trends in wind power generation: Indian scenario. *Int. J. Adv. Eng. Technol.* **2014**, *V*, 37–41.
60. World Wind Energy Association. *Worldwide Wind Capacity Reaches 744 Gigawatts—An Unprecedented 93 Gigawatts Added in 2020*; Press Release, Statistics; WWEA: Bonn, Germany, 2021.
61. Riegler, H. HAWT versus VAWT: Small VAWTs find a clear niche. *Refocus* **2003**, *4*, 44–46.
62. Kjellin, J.; Bülow, F.; Eriksson, S.; Deglaire, P.; Leijon, M.; Bernhoff, H. Power coefficient measurement on a 12 kW straight bladed vertical axis wind turbine. *Renew. Energy* **2011**, *36*, 3050–3053. [CrossRef]
63. Islam, M.; Ting, D.S.-K.; Fartaj, A. Aerodynamic models for Darrieus-type straight-bladed vertical axis wind turbines. *Renew. Sustain. Energy Rev.* **2008**, *12*, 1087–1109. [CrossRef]
64. Wahl, M. Designing an H-Rotor Type Wind Turbine for Operation on Amundsen-Scott South Pole Station. Master's Thesis, Uppsala University, Uppsala, Sweden, 2007.
65. Abdelhameed, E.H.; Hassan, H.A. Adaptive maximum power tracking control technique for wind energy conversion systems. In Proceedings of the 2018 Twentieth International Middle East Power Systems Conference (MEPCON), Nasr City, Egypt, 18–20 December 2018; pp. 146–151.
66. Hlaing, S. Basic concepts of doubly fed induction generator driven by wind energy conversion system. *Int. J. Sci. Eng. Technol. Res.* **2014**, *3*, 3242–3246.
67. Zohra, A.F.; Khalil, B.I.; Slimane, L.; Youcef, S.; Benyounes, M. Artificial intelligence control applied in wind energy conversion system. *Int. J. Power Electr. Drive Syst.* **2018**, *9*, 571.
68. Jabir, M.; Illias, H.A.; Raza, S.; Mokhlis, H. Intermittent smoothing approaches for wind power output: A review. *Energies* **2017**, *10*, 1572. [CrossRef]
69. Ding, K.; Zhi, J. *Wind Power Peak-Valley Regulation and Frequency Control Technology*; Elsevier: Amsterdam, The Netherlands, 2016; pp. 211–232.
70. Rehman, S.; Al-Hadhrami, L.M.; Alam, M.M. Pumped hydro energy storage system: A technological review. *Renew. Sustain. Energy Rev.* **2015**, *44*, 586–598. [CrossRef]
71. Rathore, N.; Yettou, F.; Gama, A. Improvement in wind energy sector using nanotechnology. In Proceedings of the 2020 6th International Symposium on New and Renewable Energy (SIENR), Ghardaia, Algeria, 13–14 October 2021; pp. 1–5.
72. Ahmed, M.; Khan, W.A. Improving Wind Turbine Performance Using Nano Materials, pp. 1–5. Available online: https://superior.edu.pk/wp-content/uploads/2017/04/13.pdf (accessed on 1 December 2021).
73. Buyuknalcaci, F.; Polat, Y.; Negawo, T.; Döner, E.; Alam, M.; Hamouda, T.; Kilic, A. Carbon nanotube-based nanocomposites for wind turbine applications. In *Polymer-Based Nanocomposites for Energy and Environmental Applications*; Elsevier: Amsterdam, The Netherlands, 2018; pp. 635–661.
74. Laturkar, S.; Mahanwar, P. Superhydrophobic coatings using nanomaterials for anti-frost applications—Review. *Nanosyst. Phys. Chem. Math.* **2016**, *7*, 650–656. [CrossRef]
75. Liu, B.; Xu, M.; Wang, L.; Tao, H.; Su, Y.; Gao, D.; Lan, L.; Zou, J.; Peng, J. Very-high color rendering index hybrid white organic light-emitting diodes with double emitting nanolayers. *Nano-Micro Lett.* **2014**, *6*, 335–339. [CrossRef]
76. Bonu, V.; Jeevitha, M.; Kumar, V.P.; Srinivas, G.; Siju; Barshilia, H.C. Solid particle erosion and corrosion resistance performance of nanolayered multilayered Ti/TiN and TiAl/TiAlN coatings deposited on Ti6Al4V substrates. *Surf. Coat. Technol.* **2020**, *387*, 125531. [CrossRef]
77. Torres-Madroñero, J.L.; Alvarez-Montoya, J.; Restrepo-Montoya, D.; Tamayo-Avendaño, J.M.; Nieto-Londoño, C.; Sierra-Pérez, J. Technological and operational aspects that limit small wind turbines performance. *Energies* **2020**, *13*, 6123. [CrossRef]
78. Cherubini, A.; Papini, A.; Vertechy, R.; Fontana, M. Airborne wind energy systems: A review of the technologies. *Renew. Sustain. Energy Rev.* **2015**, *51*, 1461–1476. [CrossRef]
79. Bechtle, P.; Schelbergen, M.; Schmehl, R.; Zillmann, U.; Watson, S. Airborne wind energy resource analysis. *Renew. Energy* **2019**, *141*, 1103–1116. [CrossRef]
80. Wisatesajja, W.; Roynarin, W.; Intholo, D. Comparing the effect of rotor tilt angle on performance of floating offshore and fixed tower wind turbines. *J. Sustain. Dev.* **2019**, *12*, 84–95. [CrossRef]
81. Taboada, J.V. Comparative analysis review on floating offshore wind foundations (FOWF). In Proceedings of the 54th Naval Engineering and Maritime Industry Congress, Ferrol, Spain, 14–16 October 2015; pp. 14–16.
82. Jamieson, P. *Innovation in Wind Turbine Design*; John Wiley & Sons: Hoboken, NJ, USA, 2018.
83. Watson, S.; Moro, A.; Reis, V.; Baniotopoulos, C.; Barth, S.; Bartoli, G.; Bauer, F.; Boelman, E.; Bosse, D.; Cherubini, A.; et al. Future emerging technologies in the wind power sector: A European perspective. *Renew. Sustain. Energy Rev.* **2019**, *113*, 109270. [CrossRef]
84. Tavner, P.; Xiang, J.; Spinato, F. Reliability analysis for wind turbines. *Wind Energy Int. J. Prog. Appl. Wind Power Convers. Technol.* **2007**, *10*, 1–18. [CrossRef]
85. Thangavelu, S.K.; Wan, T.G.L.; Piraiarasi, C. Flow simulations of modified diffuser augmented wind turbine. *IOP Conf. Ser. Mater. Sci. Eng.* **2020**, *886*, 012023. [CrossRef]

86. Ohya, Y.; Karasudani, T.; Nagai, T.; Watanabe, K. Wind lens technology and its application to wind and water turbine and beyon *Renew. Energy Environ. Sustain.* **2017**, *2*, 2. [CrossRef]
87. Razdan, P.; Garrett, P. *Life Cycle Assessment of Electricity Production from an Onshore V136-3.45 MW Wind Plant*; Vestas Wind Systen A/S: Aarhus, Denmark, 2017.
88. Robinson, L. Multi-society publication underscores MSE role in securing energy solutions. *JOM* **2010**, *62*, 96. [CrossRef]
89. Haapala, K.R.; Prempreeda, P. Comparative life cycle assessment of 2.0 MW wind turbines. *Int. J. Sustain. Manufact.* **2014**, 170–185. [CrossRef]
90. Giurco, D.; Dominish, E.; Florin, N.; Watari, T.; McLellan, B. Requirements for minerals and metals for 100% renewable scenaric In *Achieving the Paris Climate Agreement Goals*; Springer: Cham, Switzerland, 2019; pp. 437–457.
91. Oh, S.J.; Han, H.J.; Han, S.B.; Lee, J.Y.; Chun, W.G. Development of a tree-shaped wind power system using piezoelectric materia *Int. J. Energy Res.* **2010**, *34*, 431–437. [CrossRef]

materials MDPI

Article

Transport Mechanisms and Dielectric Features of Mg-Doped ZnO Nanocrystals for Device Applications

Chayma Abed [1], Amira Ben Gouider Trabelsi [2,*], Fatemah. H. Alkallas [2], Susana Fernandez [3] and Habib Elhouichet [4,5]

1 Physics Department, Faculty of Sciences of Tunis, University of Tunis El Manar, Tunis 2092, Tunisia
2 Department of Physics, College of Science, Princess Nourah Bint Abdulrahman University, P.O. Box 84428, Riyadh 11671, Saudi Arabia
3 Renewable Energy Department, CIEMAT, 28040 Madrid, Spain
4 Physics Department, College of Sciences, University of Bisha, P.O. Box 551, Bisha 61922, Saudi Arabia
5 Laboratory of Characterizations, Applications and Modelisation of Materials LR18ES08, Sciences Faculty of Tunis, University of Tunis El Manar, Tunis 2092, Tunisia
* Correspondence: aatrabelsi@pnu.edu.sa; Tel.: +966-557-570-310

Abstract: Magnesium-doped zinc oxide "ZnO:Mg" nanocrystals (NCs) were fabricated using a sol gel method. The Mg concentration impact on the structural, morphological, electrical, and dielectric characteristics of ZnO:Mg NCs were inspected. X-ray diffraction (XRD) patterns display the hexagonal wurtzite structure without any additional phase. TEM images revealed the nanometric size of the particles with a spherical-like shape. The electrical conductivity of the ZnO NCs, thermally activated, was found to be dependent on the Mg content. The impedance spectra were represented via a corresponding circuit formed by a resistor and constant phase element (CPE). A non-Debye type relaxation was located through the analyses of the complex impedance. The conductivity diminished with the incorporation of the Mg element. The AC conductivity is reduced by raising the temperature. Its plot obeys the Arrhenius law demonstrating a single activation energy during the conduction process. The complex impedance highlighted the existence of a Debye-type dielectric dispersion. The various ZnO:Mg samples demonstrate high values of dielectric constant with small dielectric losses for both medium and high-frequency regions. Interestingly, the Mg doping with 3% content exhibits colossal dielectric constant (more than 2×10^4) over wide temperature and frequency ranges, with Debye-like relaxation. The study of the electrical modulus versus the frequency and at different temperatures confirms the non-Debye relaxation. The obtained results reveal the importance of the ZnO:Mg NCs for device applications. This encourages their application in energy storage.

Keywords: ZnO nanocrystals; Mg doping; polaron hopping; high-dielectric constant; modulus

1. Introduction

Zinc oxide (ZnO) has been the subject of rapid development in the past decades in order to meet different research requirements. Essentially, its low-cost eco-friendly nature is behind its success. This explains the significant demand of ZnO material in several technological fields such as: solar cells [1], gas sensing [2], and light-emitting devices [3]. Indeed, ZnO material owns a wide band gap energy (≈3.37 eV) as well as a broad extinction binding energy (≈60 meV) at room temperature [4,5]. It is also distinguished with its notable dielectric characteristics, particularly, an elevated dielectric constant and a minor dielectric loss [6].

In this regard, the dielectric properties of ZnO are extremely attractive to researchers. Previous studies reveal the importance of developing materials with a high dielectric constant that ensures better energy storage in devices. Such materials have been widely applied to modern wireless communication technology [7] and electronic components such as capacitors. Similarly, fabricating material with low dielectric loss guarantees proper energy storage. In fact, low dielectric loss is required for non-linear optics domains, particularly at elevated frequencies [8].

ZnO materials have been enormously improved when associated with the appropriate oxides [9–11], 2D materials [12], and/or simple doping [13]. Herein, the properties of ZnO become highly sensitive to the nature of the dopant element and its concentration [14,15]. Its electrical properties depend on the excess of zinc at the interstitial position [16] and are affected essentially by the dopant element, such as, Fe [17], Al [18], and Sn [19]. However, its dielectric features drastically change with an element incorporated into its lattice [20–22]. Particularly, Mg was used to improve both the dielectric properties and the electrical conduction of ZnO. Indeed, the Zn substitution with the Mg element does not affect the lattice constants [23]. This is due to the high similarity between both elements, where their ionic radius is close to each other (i.e., Mg^{2+} = 0.057 and Zn^{2+} = 0.060). This results in a decrease in the defect's density in the ZnO lattice through a specific amount of Mg doping which ameliorates the electrical and dielectric properties of ZnO.

Deep investigations are still needed into the structural, transport, dielectric, and modulus properties of ZnO NCs-doped Mg via a low-cost technique. Earlier studies confirm the encouraging performances of the electrical and dielectric properties of ZnO:Mg in numerous applications. O. Hafef et al. [24] demonstrated the decrease in AC conductivity with increasing MgO concentration in ZnO/MgO composites. This has been assigned to the sintering technique where ions scattering form deep donor levels in the band gap. However, other studies exhibited the diminution of the AC conductivity of the material for low Mg-doping amounts [25], while high-doping levels elevate the AC conductivity of the ZnO matrix [26]. Justin Raj et al. [27] succeeded in synthesizing ZnMgO ceramics with a colossal dielectric constant that emerged from grain defects and their boundary. This reduces the conductivity relaxation due to the adsorbed water molecules. Indeed, the adsorbed moisture in the porous structure of ZnO ceramics codoped with Li and Mg contribute to the anomalous dielectric constant [28]. A similar tendency of the dielectric constant was reported for porous ZnO ceramics treated at high pressure. Here, the changes between grains and their boundary resistance was owed to a relaxation mechanism of the Maxwell–Wagner type that causes such behavior [29].

In our previous investigations, we inspected the opto-structural properties of ZnO:Mg nanocrystals (NCs) [23]. Herein, by varying the Mg content in the ZnO lattice, we expect a colossal enhancement of the dielectric constant, which is beneficial for energy storage applications. The study covers a deep investigation of the transport properties, dielectric losses, dielectric dispersion, and modulus. The impedance spectroscopy, as a nondestructive and powerful method for electrical characterization, is used to study the transport changes in both grains and grain boundaries.

2. Experimental Procedure

2.1. Synthesis of ZnO:Mg NCs

The ZnO:Mg NCs preparation technique was provided in our earlier study [23]. Briefly, zinc acetate dehydrate "$Zn(CH_3COO)_2$. $2H_2O$; 0.02M" and magnesium acetate tetrahydrate "Mg $(CH_3COO)_2$. $4H_2O$", with a purity of 99.99%, were used as starting materials. Citric acid ($C_6H_8O_7$) was used as a stabilizer and deionized water was used as a solvent. The synthesis process could be resumed as follows: 400 mL of the solvent was used to dissolve the precursors. The mixture was stirred for 2 h at 50 °C and pre-heated at 80 °C until the solvent was completely removed. Then, we proceeded to the second thermal treatment at 350 °C for 3 h. The obtained powders were crushed and annealed at 500 °C for another 3 h to achieve the crystallized phase. The thermal process was performed in two steps in order to ensure well-organized crystalline phases and reduce the amount of impurity in each sample. The prepared powders are denominated as ZnO:Mg x% where x = 0, 1, 2, 3 and 5, representing the percentage of [Mg]/[Zn].The uniaxial press of 10 tons/cm^2 was used to press the crystallized powders in order to obtain disk pellets of

1 cm diameter and a thickness "*e*" of 2 mm. The fabricated pellets were investigated during impedance measurements.

2.2. Experimental Study

XRD measurements were accomplished with a Philips X'Pert system (Malvern, UK), using a copper X-ray tube (λ = 1.54056 A°) at room temperature, at 40 kV and 100 mA. The obtained diffractometer in the angle domain was between 30° to 70° with a step size of 0.02°. The transmission electron microscopy (TEM) images were performed with a Philips CM30 microscope (Malvern, UK). The pellets were calcined and put among two platinum electrodes in a furnace. We proceeded to take the measurements by varying the temperature for each sample with a step of 10 °C from 200 to 260 °C. The study of the dielectric properties at different temperatures was conducted by gathering complex impedance data in the frequency domain between 40 Hz to 5 MHz with an impedance analyzer (Agilent 4294A, USA).

3. Results and Discussion

3.1. Structural and Morphological Studies

Figure 1 represents the XRD patterns of pure and Mg-doped ZnO NCs. The obtained diffraction peaks highlighted the hexagonal wurtzite structure of ZnO of the different samples. The patterns demonstrate high crystalline quality along the (101) plane as a preferred orientation.

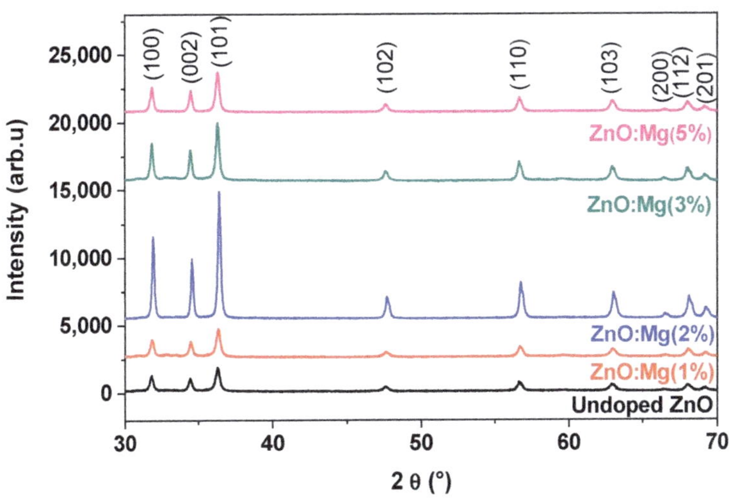

Figure 1. X-ray diffraction patterns of the prepared ZnO:Mg NCs.

The ZnO structure remains unchanged after the Mg incorporation with low content. This affects the intensity and the wideness of the peaks. Indeed, the full width at half maximum of the peaks along the (101) plane decreased with Mg doping. The FWHM of the doped sample is reduced compared with that of a pure ZnO, highlighting the improvement of the crystalline structure. On other hand, the crystallite size increases with the incorporation of Mg, which was found to be greater than that of the undoped sample. Such an enhancement of the crystallite size is assigned to the Zn^{2+} substitution with Mg^{2+}. The obtained values are gathered in Table 1.

Table 1. Structural parameters of pure ZnO and ZnO:Mg NCs [23].

	FWHM of (101) Peak (Rad)	D (nm)
ZnO:Mg 0%	0.0060	26.82
ZnO:Mg 1%	0.0054	30.47
ZnO:Mg 2%	0.0038	42.96
ZnO:Mg 3%	0.0044	36.75
ZnO:Mg 5%	0.0048	33.69

The TEM images of non-doped ZnO and ZnO:Mg (3%) NCs are given in Figure 2. A homogenous structure with a spherical shape could be observed on the fabricated NCs. The mean particle size is about 30 nm, which is in good agreement with the XRD analysis. Moreover, it is clear that the nanocrystal shape remains the same after doping. However, it is difficult to judge if the size decreases with Mg doping.

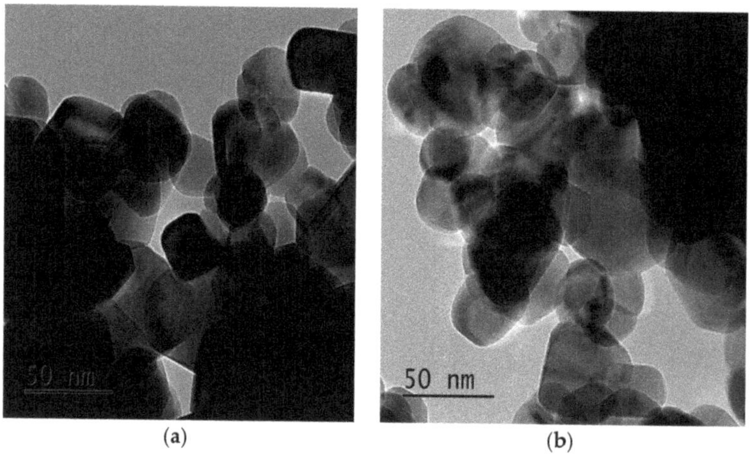

Figure 2. TEM images of undoped (**a**) and Mg3%-doped (**b**) ZnO NCs.

3.2. Impedance Spectroscopy

3.2.1. Electrical Study

a. Impedance Spectra

The complex impedance plots of undoped ZnO at various temperatures is given in Figure 3. We also studied the complex impedance changes at 200 °C for the various Mg percentages (see, Figure 4). Both figures illustrate the well-defined semicircles in the complex plane. This confirms the homogeneity and the single phase of the prepared doped samples. The centers of the semicircles are localized below the Z' axis, which may be related to a non-Debye relaxation nature [30,31]. Compared with pure ZnO, the semicircles of Mg-doped ZnO NCs appeared at higher temperatures. Here, the increase in the Mg concentration enlarges the semicircle radius. However, the sample doped with 3% of Mg displays an opposite tendency. The diminution in the radius of the semicircle illustrates the enhanced of conductivity in the sample [32].

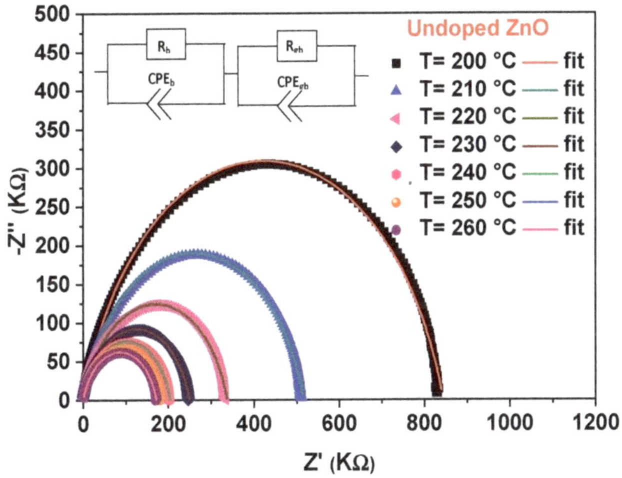

Figure 3. Nyquist plots of the undoped ZnO.

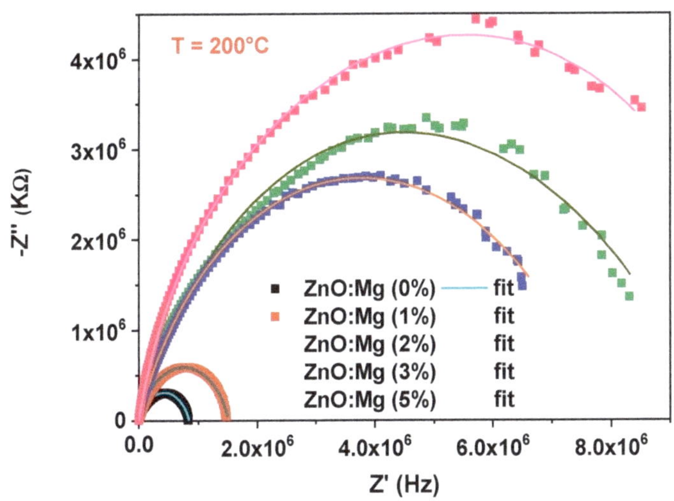

Figure 4. Experimental and theoretical Nyquist plots of ZnO:Mg NCs at T = 200 °C.

The investigation of the impedance plot requires modeling of the electric impedance with a corresponding circuit including one block consisting of a parallel association of the bulk resistance R and a capacitor.

In this case, the capacitor could be substituted by a constant phase element (CPE) Z_{CPE} because of the semicircles depression [33]. Z_{CPE} is given by:

$$Z_{CPE} = \frac{1}{A_0(j\omega)^n} \tag{1}$$

Here, j denotes the imaginary unit, ω corresponds to the angular frequency, A_0 is the capacitance value of Z_{CPE}, and n is non-dimensional parameter ($0 \leq n \leq 1$) reflecting the trend to an exact semicircle [34].

The total impedance Z^* is described as follows:

$$Z = Z' - jZ'' \tag{2}$$

where

$$Z' = \frac{R \times \left(1 + R \times A_0 \times (2\pi x)^n \times \cos\left(\frac{\pi}{2}n\right)\right)}{1 + 2 \times R \times A_0 \times (2\pi x)^n \times \cos\left(\frac{\pi}{2}n\right) + \left(R \times A_0 \times (2\pi x)^n\right)^2} \quad (3)$$

$$-Z'' = \frac{R^2 \times A_0 \times (2\pi x)^n \times \sin\left(\frac{\pi}{2}n\right)}{1 + 2 \times R \times A_0 \times (2\pi x)^n \times \cos\left(\frac{\pi}{2}n\right) + \left(R \times A_0 \times (2\pi x)^n\right)^2} \quad (4)$$

The suggested model is used to fit the Nyquist plots, proving the viability between the equivalent circuit and the existing system (see Figures 3 and 4). Table 1 lists the extract parameters. The capacitances, A_g, is sized up at 10^{-10} F cm^{-2} S^{n-1}, demonstrating that the grain boundary contribution is formed as a general response of the system [35,36]. The slight difference in ionic radius between Zn and Mg can generate a small displacement in the ZnO structure, since the Mg substitutes the interstitial Zn sites [23]. The n_g exponent values remain unaffected with increasing the Mg content oppositely to the resistance R_g that raises in this case (see, Table 2). This may be explained by the structural defects related to the doping. A similar behavior was observed for ZnO-doped Sb [37], where the enhancement of the resistance is coming from the substitution of the Zn^{2+} ions with Sb^{3+}. Previous work by M. Ben Ali et al. [38] examining the Ni-doping influence on the electrical properties of ZTO assigned the increase in resistivity rate to the placement of the excess of the Ni-doping element on the grain boundaries. This demonstrates that Ni atoms are non-localized in the ZTO matrix, where the Ni excess leads to the formation of a Schottky barrier blocking the carrier's transport. Based on these results, we can conclude that for high-doping concentrations, Mg atoms prefer to sit in the grain boundaries and form a barrier to hinder the carrier's transport.

Table 2. Fitting parameters relative to the equivalent circuit elements for all the ZnO:Mg samples.

	R_b (10^5 Ω)	A_g (10^{-10} F cm^{-2} S^{n-1})	n_g
Mg 0%	1.82848	1.9535	0.8023
Mg 1%	8.17662	1.7426	0.8325
Mg 2%	35.72726	2.3201	0.7633
Mg 3%	30.83072	2.4610	0.7645
Mg 5%	49.20645	1.6784	0.8153

b. DC Conductivity

In order to inspect the electrical properties (DC conductivity, σ_{DC}) of the Mg-doped ZnO NCs, we use the following expression:

$$\sigma_{DC} = \frac{e}{SZ_0} \quad (5)$$

where e is the sample thickness, S represents the surface region, and Z_0 characterizes the resistance found based on the interception of the semicircles with the real axis.

The reciprocal temperature dependence to the DC conductivity is drawn in straight lines in Figure 5. Here, the Arrhenius law could be used:

$$\sigma_{DC}T = \sigma_0\ .\exp\left(-\frac{E_a}{K_B.T}\right) \quad (6)$$

where, K_B designed the Boltzmann constant and E_a is the activation energy corresponding to the energy difference between the conduction levels and the donor. The E_a values are deduced from the linear fit slope of the logarithmic DC conductivity $(\sigma_{dc}.T)$ versus $\frac{1000}{T}$. We found low activation energy values of doped samples in comparison with unmodified ZnO (see Table 3). The activation energy is estimated at 0.25 eV for the 1% Mg-doping concentration, which is ascribed to Zn^{2+} [39]. Besides, such an activation energy is found in the range (0.32 to 0.38 eV) for higher Mg percentages. These values originate from the oxygen vacancies V_O [40]. Chaari et al. [19] reported comparable results for ZnO ceramics doped with high concentrations of Sn_2O_3. The lower activation energy values reveal the source of the conduction phenomenon process in ZnO:Mg NCs, which is associated with the polaron hopping.

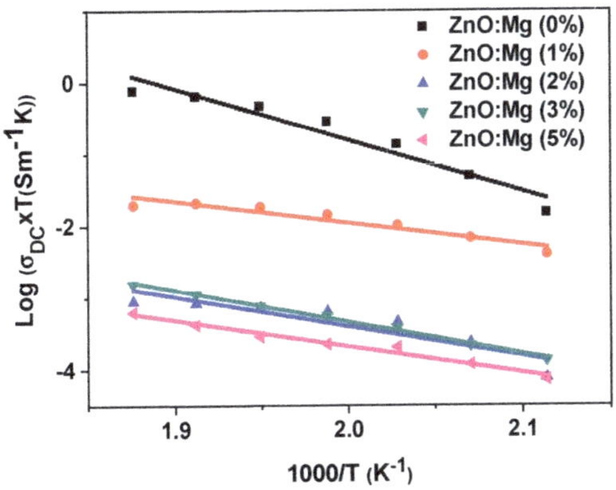

Figure 5. Semi-logarithmic plots of (σ_{DC}T) versus 1000/T according to Arrhenius law.

Table 3. σ_{dc} and E_a changes with Mg content at T = 250 °C.

% Mg	0	1	2	3	5
R_b (kΩ)	840.56	1481.82	6589.22	8321.14	8424.57
$\sigma_{DC}(10^{-6}$ S.m$^{-1})$	15.15	8.56	1.93	1.53	1.51
E_a (eV)	0.62	0.25	0.36	0.38	0.32

The DC conductivity decreases significantly with increasing the Mg doping amount due to the diminishing of the free electrons density. In fact, the imperfections act as trapping/scattering centers to reduce the number of free electrons. Further, the Mg atoms located at the grain boundaries behave as an electrical barrier that raise the carrier scattering, therefore, decreasing the conductivity [19,41].

c. AC conductivity

Figure 6a represents the frequency–temperature dependency of the AC conductivity. This is demonstrated by the Jonsher's relationship [38,42]:

$$\sigma_{AC}(\omega) = \sigma_{dc} + A\omega^s \quad (7)$$

where, σ_{dc} is the DC conductivity associated to the flat region, $A\omega^s$ is the frequency reliant expression that characterizes the dispersion phenomenon, A a constant associated with the strength of polarizability [42], and s is a factor verifying $0 \leq s \leq 1$ [43] which describes the interaction level between the mobile ions and the matrix. All these parameters depend on temperature [44].

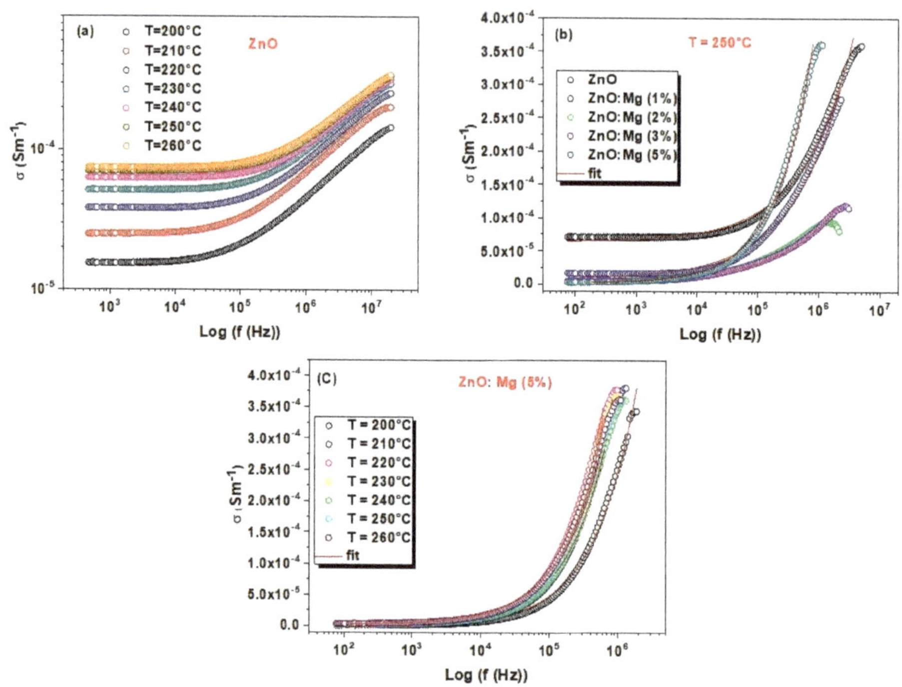

Figure 6. Frequency dependance of σ_{AC} for: (**a**) undoped ZnO at different temperatures (**b**) all the ZnO:Mg samples at T = 250 °C and (c) ZnMg5 at different temperatures.

As we can see from Figure 6a, all the curves depend on temperature. The power law is obeyed for the different ZnO:Mg samples since the fit matches well with the experimental data in Figure 6a–c. At low-frequency range, the plots show an unaffected flat region corresponding to DC conductivity, whereas the dispersion at high-frequency range is related to AC conductivity. Ionic conductors are characterized with such a behavior [45]. For the lower frequencies, the conductivity is reduced with further Mg addition due to the placement of the doping element on the interstitial position in the ZnO lattice that enhances structural defects concentration (Figure 6b). At the high-frequency region, the conductivity continues to decrease from 1 to 3% of Mg content while it shows an increase for 5% of Mg doping. These latest samples display a high conductivity due to the increase in oxygen vacancies that lead to augmentation of the hopping charge carrier's concentration [46].

The theoretical fits give σ_{dc} and s values which can explain the mechanism related to the conduction process. The s parameter is dependent on the ZnO and ZnO:Mg (1%) temperature, shown in Figure 7. It slightly increases from 0.42 to 0.56 with temperature. This increase could be attributed to the quantum mechanical tunneling (QMT) conduction model [47]. Here, s is expressed as [48]:

$$s = 1 - \frac{4}{\ln\left(\frac{1}{\omega\tau_0}\right) - \frac{W_H}{K_B T}} \tag{8}$$

where, K_B is the Boltzmann constant, W_H is the energy of the polaron hopping, and τ_0 is a typical relaxation time. When $W_H >> K_B T$, the expression can be simplified as follows:

$$s = 1 + \frac{4k_B T}{W_H} \tag{9}$$

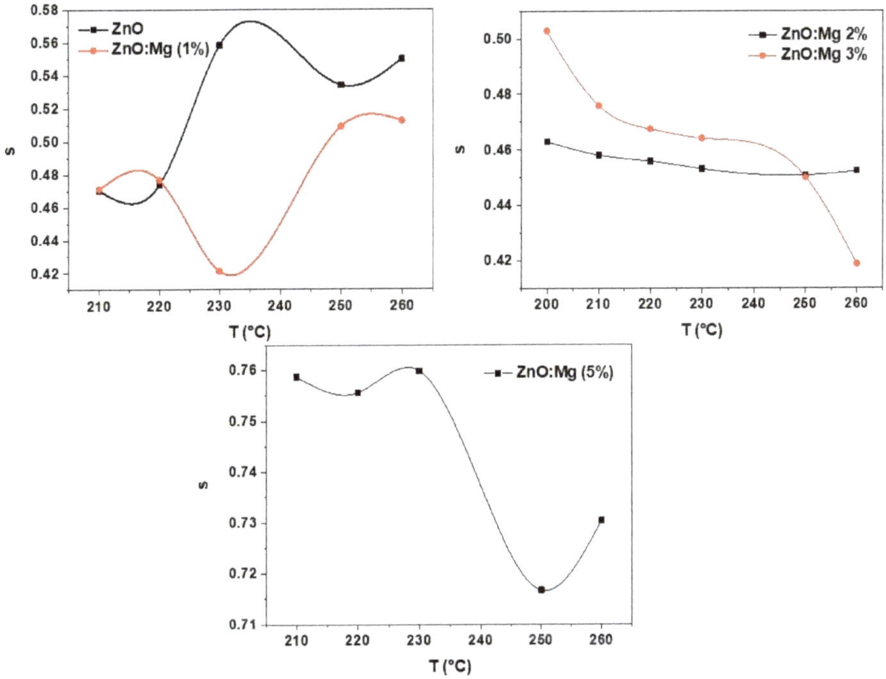

Figure 7. Temperature dependence of the parameter s for all the ZnO:Mg samples.

For the samples ZnM2 and ZnMg3, the exponent s was found to vary from 0.50 to 0.42 with increasing temperature from 200 to 260 °C. However, its value is more important for the ZnO:Mg (5%) sample, around 0.7, and decreases slightly with temperature. The diminution of the s parameter leads to the principal transfer route which is correlated with the barrier hopping (CBH) model [49–51], where charge carriers skip among sites over the potential barrier parting them instead of tunneling via the barrier [52]. The parameter s obeys the following equation [53]:

$$s = 1 - \frac{6K_B T}{W_M} \tag{10}$$

W_M is the upper barrier height at an infinite separation. This is also identified as the energy relative to polaron binding in its localized sites.

Figure 8 displays the (1-s) plots versus the temperature that gives the W_M values . W_M and s raised with Mg concentration (see Figure 8). Such a change in conduction model is assigned to the enhancement of the charge carrier induced by more Mg^{2+} ions in the ZnO lattice.

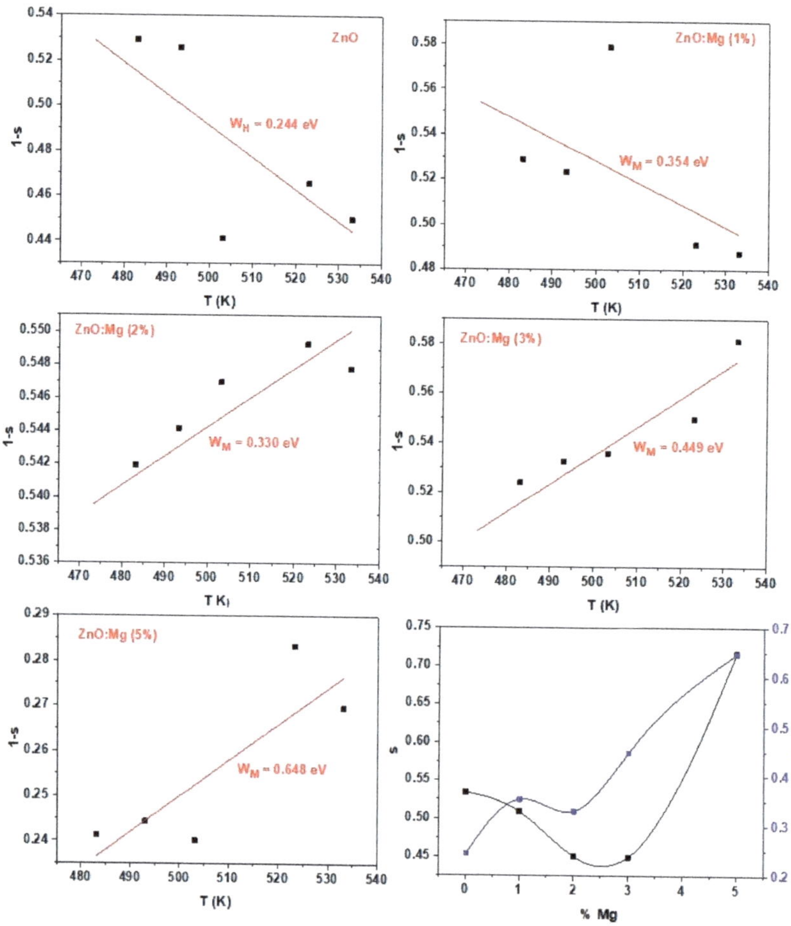

Figure 8. Variations of the parameters (1-s), and W_M with temperature, and %Mg, respectively.

3.2.2. Dielectric Study

a. Permittivity and loss studies

The complex dielectric constants describe the dielectric properties of materials. They are determined from Z' and Z'' (Equations (3) and (4), respectively) and given by the following formula [38]:

$$\varepsilon(\omega) = \varepsilon'(\omega) - j\varepsilon''(\omega) = \frac{1}{j\omega C_0(Z' + jZ'')} \tag{11}$$

where the real ε' and the imaginary ε'' parts illustrate the quantities of energy accumulated and dispersed in the dielectric owing to the applied electric field, correspondingly. Their expressions are:

$$\varepsilon'(\omega) = -\frac{Z''}{\omega C_0\left(Z'^2 + Z''^2\right)} \tag{12}$$

$$\varepsilon''(\omega) = \frac{Z'}{\omega C_0\left(Z'^2 + Z''^2\right)} \tag{13}$$

Here, C_0 is geometrical capacitance of samples ($C_0 = \frac{\varepsilon_0 S}{e}$, ε_0 is the permittivity of the vacuum, S is the cross-sectional area of the flat surface of the pellet, and e is the thickness). Z' and Z'' are the real and the imaginary parts of impedance. The dielectric loss ($\tan\,\delta$) is calculated from the values of the dielectric constants and could be written as follows [54]:

$$\tan\,\delta = \frac{\varepsilon''}{\varepsilon'} \tag{14}$$

Figure 9a depicts the frequency dispersion ε' for all the ZnO:Mg samples. It shows an increase of ε' with the Mg content. For each sample, the dielectric constant gradually reduces as the frequency augments, and reaches an almost constant value (relatively high) in the high frequencies range.

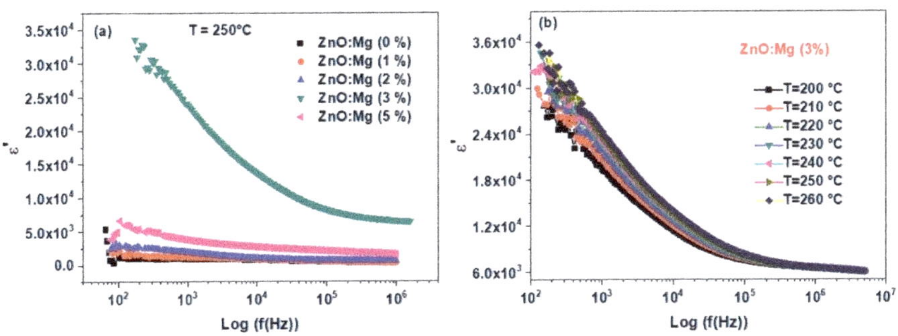

Figure 9. Dielectric constant ε' versus frequency for (**a**) different Mg doping and (**b**) ZnO:Mg (3%) sample, at various temperatures.

Figure 9b depicts the frequency dispersion ε' versus the temperature of the ZnO:Mg (3%) sample. As shown, ε' is gradually reduced through the increasing of the frequency to reach almost a constant value in the high frequencies range. This way points to a Debye-type dielectric dispersion [55,56]. According to the Maxwell–Wagner model [57], the large values of ε' is related to the large polarization effect produced at the grain boundaries due to the charge carrier's migration between grain boundaries under the effect of the applied external field [58]. At high frequencies range, the decrease in ε' values could be explained by polarizability loss, since the dipoles are enabled to rotate quickly which leads to a delay among frequencies of the oscillating dipole and applied fields [59]. It can be pointed out that the reached ε' value (at medium and high frequencies) is considered relatively high, making the ZnO:Mg samples very promising for energy storage applications.

The imaginary part ε'' also decreases with the frequency but it varies slightly versus the temperature. This demonstrates the loss of polarization with the frequency by the disappearance of dipoles or to their inability to rotate properly at high frequencies [60].

We located a considerable reduction of the imaginary part ε'' with the rate of the Mg doping which confirms the rule of the space charge [61] (see Figure 10a). The same performance was also monitored by Hafef et al. [24] with high MgO-doping rates (10% and 20% of MgO). This also was explained by the separation of the MgO phase in the ZnO host matrix. In this work, the Mg is well incorporated into the ZnO lattice and hence, it induces less crystallinity, as was previously shown in the XRD analysis. Thus, this behavior could be related to the increased defects density induced by further Mg incorporation, which leads to less dielectric polarization. Moreover, ε'' increases with the temperature (see Figure 10b) where no peaks in the loss tangent curves are observed (Figure 11). This demonstrates the slightly high conduction losses [62]. Finally, the present material could be an encouraging candidate for device applications acting in the high-frequency range.

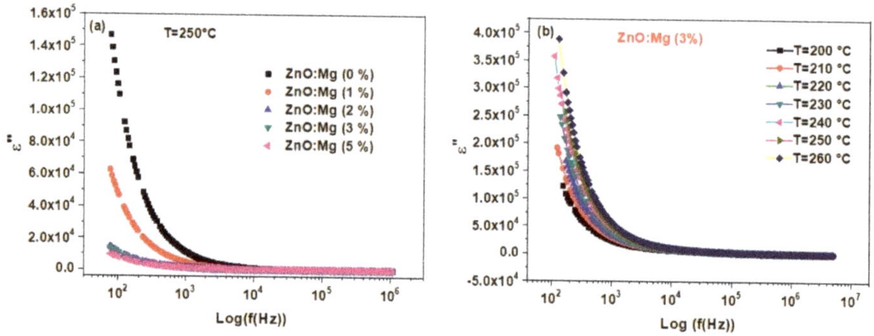

Figure 10. Frequency dependence curves of the imaginary part ε'' (**a**) for the ZnO:Mg (3%) sample at different temperatures and (**b**) for different Mg content.

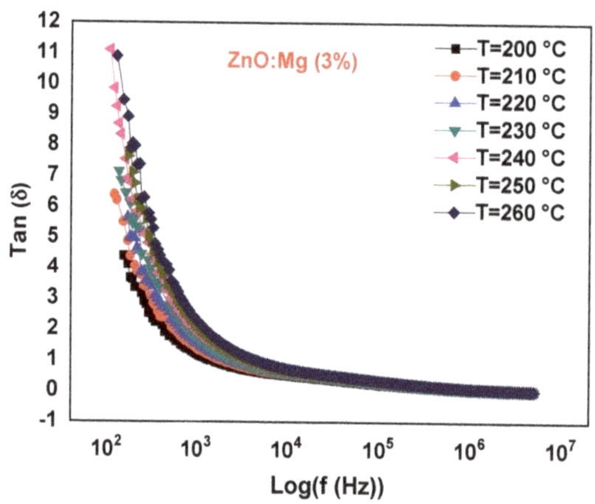

Figure 11. Dependance of the dielectric loss tan δ on the frequency for the sample ZnO:Mg 3% at different temperatures.

b. Modulus analysis

The electric modulus is generally used to examine the electrical characteristics of bulk materials. At low frequencies, the conductivity phenomenon usually hides the interfacial polarization generally present in such materials [63]. To surmount this difficulty, modulus is mostly appropriate to obtain phenomena such as the conductivity relaxation times and the electrode polarization. The complex electric modulus is written as [64]:

$$M^* = \frac{1}{\varepsilon^*(\omega)} = M'(\omega) + jM''(\omega) = j\omega C_0 Z^* \tag{15}$$

where, $M' = \omega C_0 Z''$ and $M'' = \omega C_0 Z'$ are the real and imaginary parts of the complex electric modulus, respectively.

Figure 12a represents the variations of M' within the frequency at different temperatures. It raises with increasing frequency to reach a flat zone to the highest frequency values corresponding to the limiting value of M'. This behavior illustrates the weak contribution of the electrode polarization that could be neglected for Mg-doped ZnO NCs [65]. Such behavior was also detected for Na-doped ZnO NCs [22].

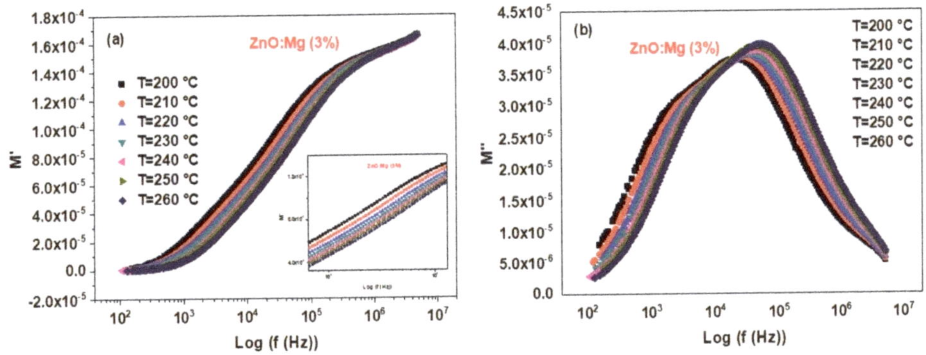

Figure 12. Plots of (**a**) M' and (**b**) M'' within frequency for various temperatures for the ZnO:Mg (3%) sample.

The frequency reliance of M'' with the temperature for ZnO:Mg (3%) is represented in Figure 12b. The M''_{max} position shifts near the high-frequency region when the temperature increases, and the shift of the M''_{max} position is usually related to the conductivity relaxation. An asymmetric peak could be observed for each investigated temperature, which exhibited the non-Debye kind of behavior in the relaxation of the Mg-doped ZnO NCs [66].

The detected peak aids in determining the relaxation time τ using the following relationship [58]:

$$\tau = \frac{1}{\omega_{max}} = \frac{1}{2\pi\, f_{max}} \tag{16}$$

where f_{max} designates the relaxation frequency.

The change of M''_{max} with temperature images the enhancement of the dielectric relaxation time τ with temperature [37]. Such a change affirms the thermal stimulation of the dielectric relaxation [67]. As shown in Figure 13, $Ln\tau$ varies linearly with the inverse of the absolute temperature T, following the equation:

$$\tau = \tau_0 \exp\left(-\frac{E_a}{K_B T}\right) \tag{17}$$

where, τ_0 is the pre-exponential factor. From a linear fit of the plot, the value of the activation energy E_a is estimated to 0.73 eV, which is quite different to that determined while considering the plot of Ln(σ_{dc}T). Hence, it provides information about the non-statistic distribution of the Mg^{2+} ions in the ZnO matrix and suggests an arbitrary conductivity. Therefore, the relaxation of dipoles manifests itself arbitrarily [37,68].

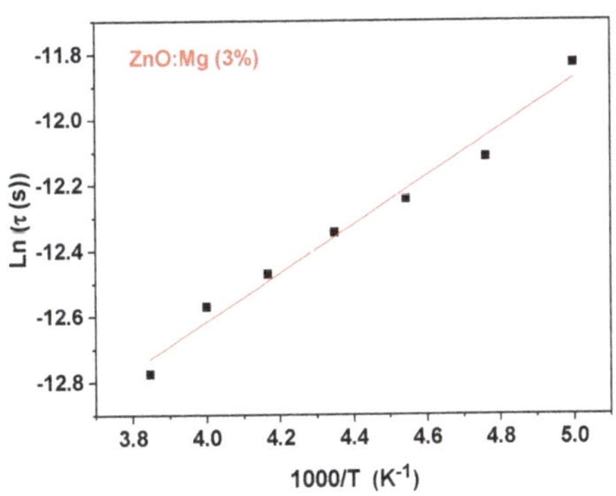

Figure 13. Plot of $Ln\ \tau$ versus 1000/T.

4. Conclusions

Mg-doped ZnO NCs were synthesized via sol-gel method. The electrical and dielectric characteristics were investigated as function of temperature and frequency, using impedance spectroscopy. The conductivity and the activation energy are influenced by Mg concentration. From temperature reliance, we reveal that the conduction mechanism is managed from the model of the hopping correlated barrier (CBH). Dielectric features are largely affected by Mg doping due to less dielectric polarization. The low dielectric losses at high frequencies and the shift toward high frequencies of M'' with temperature, make the Mg-doped ZnO NCs a suitable material for application in non-linear optics. The analysis of the electrical modulus illustrates an extension in the relaxation time with further Mg incorporation. All the dielectric properties relative to the prepared ZnO:Mg NCs are of interest to bring a considerable influence to several technological applications, such as microwave devices.

Author Contributions: Conceptualization C.A. and H.E.; investigation, C.A., A.B.G.T., F.H.A., S.F. and H.E.; formal analysis, C.A., A.B.G.T., F.H.A., and S.F.; methodology, C.A., A.B.G.T. F.H.A. and S.F.; data curation, C.A., A.B.G.T., F.H.A. and S.F.; writing—original draft preparation, C.A. and A.B.G.T.; writing—review and editing, C.A., A.B.G.T., F.H.A., S.F. and H.E.; supervision H.E. All authors have read and agreed to the published version of the manuscript.

Funding: This research was funded by Princess Nourah bint Abdulrahman University Researchers Supporting Project number (PNURSP2022R223), Princess Nourah bint Abdulrahman University, Riyadh, Saudi Arabia.

Institutional Review Board Statement: Not applicable.

Informed Consent Statement: Not applicable.

Acknowledgments: The authors express their gratitude to Princess Nourah bint Abdulrahman University Researchers Supporting Project number (PNURSP2022R223), Princess Nourah bint Abdulrahman University, Riyadh, Saudi Arabia.

Conflicts of Interest: The authors declare no conflict of interest.

References

1. Mohammed, M.K.A.; Shekargoftar, M. Surface treatment of ZnO films with carbon nanotubes for efficient and stable perovskite solar cells. *Sustain. Energy Fuels* **2021**, *5*, 540–548. [CrossRef]
2. Ying, S.; Wang, Y.; Wu, Z.; Huang, M.; Dong, L.; Zhao, J.; Peng, C. Highly-sensitive NO2 gas sensors based on three-dimensional nanotube graphene and ZnO nanospheres nanocomposite at room temperature. *Appl. Surf. Sci.* **2021**, *566*, 150720. [CrossRef]
3. Barman, B.; Swami, S.K.; Dutta, V. Fabrication of highly conducting ZnO/Ag/ZnO and AZO/Ag/AZO transparent conducting oxide layers using RF magnetron sputtering at room temperature. *Mat. Sci. Semicond. Process.* **2021**, *129*, 105801. [CrossRef]
4. Özgür, Ü.; Alivov, Y.I.; Liu, C.; Teke, A.; Reshchikov, M.; Dogan, S.; Avrutin, V.; Cho, S.-J.; Morkoc, H. A comprehensive review of ZnO materials and devices. *J. App. Phys.* **2005**, *98*, 041301. [CrossRef]
5. Baladin, A.A.; Wang, K.L. *Handbook of Semiconductor Nanostructures and Nanodevices*; American Scientific Publisher: Los Angeles, CA, USA, 2006.
6. Chaari, M.; Matoussi, A. Electrical conduction and dielectric studies of ZnO pellets. *Phys. B* **2012**, *407*, 3441–3447. [CrossRef]
7. Zhang, R.; Liu, J.; Wang, Y.; Luo, Z.; Zhang, B.; Duan, J. Flexible Wearable Composite Antennas for Global Wireless Communication Systems. *Sensors* **2021**, *21*, 6083. [CrossRef] [PubMed]
8. Kanagathara, N.; Renganathan, N.G.; Marchewka, M.K.; Sivakumar, N.; Gayathri, K.; Krishnan, P.; Gunasekaran, S.; Anbalagan, G. Spectrochimica. *Acta A* **2013**, *101*, 112–118. [CrossRef]
9. Abed, C.; Fernandez, S.; Elhouichet, H. Studies of optical properties of ZnO:MgO thin films fabricated by sputtering from home-made stable oversize targets. *Optik* **2020**, *216*, 164934. [CrossRef]
10. Singh, J.; Singh, R.C. Structural, optical, dielectric and transport properties of ball mill synthesized ZnO–V2O5 nanocomposites. *J. Mol. Struct.* **2020**, *1215*, 128261. [CrossRef]
11. Ali, M.B.; Bouaifel, F.; Sieber, B.; Elhouichet, H.; Addad, A.; Boussekey, L.; Férid, M.; Boukherroub, R. Preparation and characterization of Ni-doped ZnO-SnO2 nanocomposites: Application in photocatalysis. *Superlattices Microstruct.* **2016**, *91*, 225–237.
12. Wang, S.; Tian, H.; Ren, C.; Yu, J.; Sun, M. Electronic and optical properties of heterostructures based on transition metal dichalcogenides and graphene-like zinc oxide. *Science* **2018**, *8*, 12009. [CrossRef] [PubMed]

13. Nasser, R.; Song, J.; Elhouichet, H. Epitaxial growth and properties study of p-type doped ZnO:Sb by PLD. *Superlattices Microstruct.* **2021**, *155*, 106908. [CrossRef]
14. Othmen, W.B.H.; Ali, M.B.; Bouslama, W.; Elhouichet, H. Solar driven photocatalytic properties of Sm3+ doped ZnO nanocrystals. *Ceram. Int.* **2020**, *46*, 18878.
15. Nasser, R.; Elhouichet, H. Production of acceptor complexes in sol-gel ZnO thin films by Sb doping. *J. Lumin.* **2018**, *196*, 11–19. [CrossRef]
16. Vimalkumar, T.V.; Poornima, N.; Kartha, C.S.; Vijayakumar, K.P. Effect of precursor medium on structural, electrical and optical properties of sprayed polycrystalline ZnO thin films. *Mater. Sci. Eng. B* **2010**, *175*, 29–35. [CrossRef]
17. Saadi, H.; Rhouma, F.I.H.; Benzarti, Z.; Bougrioua, Z.; Guermazi, S.; Khirouni, K. Electrical conductivity improvement of Fe doped ZnO nanopowders. *Mater. Res. Bull.* **2020**, *129*, 110884. [CrossRef]
18. Li, Y.; Feng, J.; Wang, Y.; He, B.; Zhao, Y.; Xu, C.; Wang, J. Highly conductive and flexible electrodes based on ultrathin aluminum-doped zinc oxide epitaxial films. *Appl. Surf. Sci.* **2021**, *568*, 150925. [CrossRef]
19. Chaari, M.; Matoussi, A. Effect of Sn2O3 doping on structural, optical and dielectric properties of ZnO ceramics. *Mater. Sci. Eng. B* **2013**, *178*, 1130–1139. [CrossRef]
20. Singh, J.; Singh, R.C. Tuning of structural, optical, dielectric and transport properties of Fe-doped ZnO:V system. *Mat. Sci. Semicon. Proc.* **2021**, *121*, 105305. [CrossRef]
21. Singh, J.; Singh, R.C. Enhancement of optical, dielectric and transport properties of (Sm, V) co-doped ZnO system and structure-property correlations. *Ceram. Int.* **2021**, *47*, 10611–10627. [CrossRef]
22. Tabib, A.; Sdiri, N.; Elhouichet, H.; Férid, M. Investigations on electrical conductivity and dielectric properties of Na doped ZnO synthesized from sol gel method. *J. Alloys. Compd.* **2015**, *622*, 687–694. [CrossRef]
23. Abed, C.; Bouzidi, C.; Elhouichet, H.; Gelloz, B.; Ferid, M. Mg doping induced high structural quality of sol–gel ZnO nanocrystals: Application in photocatalysis. *Appl. Surf. Sci.* **2015**, *349*, 855–863. [CrossRef]
24. Hafef, O.; Othman, Z.J.; Megdich, M.; Matoussi, A. Conduction mechanism and dielectric properties of ZnO/MgO solid composites. *Appl. Phys. A* **2017**, *123*, 95. [CrossRef]
25. Iqbal, J.; Jan, T.; Ismail, M.; Ahmad, N.; Arif, A.; Khan, M.; Adil, M.; Haq, S.U.; Arshad, A. Influence of Mg doping level on morphology, optical, electrical properties and antibacterial activity of ZnO nanostructures. *Ceram. Int.* **2014**, *40*, 7487–7493. [CrossRef]
26. Rouchdi, M.; Salmani, E.; Fares, B.; Hassanain, N.A. Mzerd, Synthesis and characteristics of Mg doped ZnO thin films: Experimental and ab-initio study. *Results Phys.* **2017**, *7*, 620–627. [CrossRef]
27. Raj, C.J.; Paramesh, G.; Prakash, B.S.; Meher, K.R.S.P.; Varma, K.B.R. Origin of giant dielectric constant and conductivity behavior in Zn1− xMgxO ($0 \leq x \leq 0.1$) ceramics. *Mat. Res. Bul.* **2016**, *74*, 1–8. [CrossRef]
28. Soukiassian, A.; Tagantsev, A.; Setter, N. Anomalous dielectric peak in Mg and Li doped ZnO ceramics and thin films. *Appl. Phys. Lett.* **2010**, *97*, 192903. [CrossRef]
29. Li, X.; Xu, L.; Liu, L.; Wang, Y.; Cao, X.; Huang, Y.; Meng, C.; Wang, Z. High pressure treated ZnO ceramics towards giant dielectric constants. *J. Mater. Chem. A* **2014**, *2*, 16740–16745. [CrossRef]
30. Bonneau, P.; Garnier, O.; Calvarin, G.; Husson, E.; Gahvarri, J.R.; Hewat, A.W. Morrel—X-ray and neutron diffraction studies of the diffuse phase transition in PbMg13Nb23O3 ceramics. *J. Sol. State Chem.* **1991**, *91*, 350–361. [CrossRef]
31. Othmen, W.B.H.; Sdiri, N.; Elhouichet, H.; Férid, M. Study of charge transport in Fe-doped SnO2 nanoparticles prepared by hydrothermal method. *Mat. Sci. Semicon. Proc.* **2016**, *52*, 46–54. [CrossRef]
32. Palmer, G.; Poeppelmeier, K.R. Phase relations, transparency and conductivity in Ga2O3-SnO2- ZnO. *Solid State Sci.* **2002**, *4*, 317–322. [CrossRef]
33. Sdiri, N.; Elhouichet, H.; Elakermi, E.; Dhifallah, A.; Ferid, M. Structural investigation of amorphous Na2O–P2O5–B2O3 correlated with its ionic conductivity. *J. Non-Cryst. Sol.* **2015**, *409*, 34–42. [CrossRef]
34. Jorcin, J.B.; Orazem, M.E.; Pebere, N.; Tribollet, B. CPE analysis by local electrochemical impedance spectroscopy. *Electrochim. Acta* **2006**, *51*, 1473–1479. [CrossRef]
35. Landau, L.; Lifchitz, E. *Electrodynamique des Milieux Continus*; Mir: Moscow, Russia, 1969; p. 325.
36. Mahamoud, H.; Louati, B.; Hlel, F.; Guidara, K. Impedance and modulus analysis of the (Na0.6Ag0.4)2PbP2O7 compound. *J. Alloys. Compd.* **2011**, *509*, 6083–6089. [CrossRef]
37. Nasser, R.; Othmen, W.B.H.; Elhouichet, H. Effect of Sb doping on the electrical and dielectric properties of ZnO nanocrystals. *Ceram. Int.* **2019**, *45*, 8000–8007. [CrossRef]
38. Ali, M.B.; Elhouichet, H. Electrical and dielectric properties of Ni doped Zn2SnO4 nanoparticles. *Ceram. Int.* **2020**, *46*, 28686–28692.
39. Nahm, C.-W. Effect of dopant (Al, Nb, Bi, La) on varistor properties of ZnO–V2O5–MnO2–Co3O4–Dy2O3 ceramics. *Ceram. Int.* **2010**, *36*, 1109–1115. [CrossRef]
40. Hong, Y.; Kim, J. The electrical properties of Mn3O4-doped ZnO. *Ceram. Int.* **2004**, *30*, 1301–1306. [CrossRef]
41. Othman, Z.J.; Hafef, O.; Matoussi, A.; Rossi, F.; Salviati, G. AC conductivity and structural properties of Mg-doped ZnO ceramic. *Appl. Phys. A* **2015**, *121*, 625–634. [CrossRef]
42. Balasubramanian, S.; Kakimoto, K.; Ohsato, H. Frequency and temperature dependent dielectric and conductivity behavior of KNbO3 ceramics. *J. Appl. Phys.* **2003**, *94*, 5182–5187.
43. Hill, R.; Jonscher, A. DC and AC conductivity in hopping electronic systems. *J. Non-Cryst. Solids* **1979**, *32*, 53–69. [CrossRef]

44. Langar, A.; Sdiri, N.; Elhouichet, H.; Férid, M. Conductivity and dielectric behavior of NaPO3–ZnO–V2O5 glasses. *J. Alloy Compd.* **2014**, *590*, 380–387. [CrossRef]
45. Angell, C. Fast ion motion in glassy and amorphous materials. *Solid State Ion.* **1983**, *9-10*, 3–16. [CrossRef]
46. Aakansha; Deka, B.; Ravi, S.; Pamu, D. Impedance spectroscopy and ac conductivity mechanism in Sm doped Yttrium Iro Garnet. *Ceram. Int.* **2017**, *43*, 10468–10477. [CrossRef]
47. Mansour, S.A.; Yahia, I.S.; Yakuphanoglu, F. The electrical conductivity and dielectric properties of C.I. Basic Violet 10. *Dyes Pigr* **2010**, *87*, 144–148. [CrossRef]
48. Elliott, S.A.C. conduction in amorphous chalcogenide and pnictide semiconductors. *Adv. Phys.* **1987**, *36*, 135–217. [CrossRef]
49. Winie, T.; Arof, A.K. Dielectric behaviour and AC conductivity of LiCF3SO3 doped H-chitosan polymer films. *Ionics* **2004**, *1* 193–199. [CrossRef]
50. Jung, W.H. AC conduction mechanisms of Gd1/3Sr2/3FeO3 ceramic. *Phys. B Cond. Mat.* **2008**, *403*, 636–638. [CrossRef]
51. Elliott, S.R. Temperature dependence of a.c. conductivity of chalcogenide glasses. *Philos. Mag. Part B* **1978**, *37*, 553–560. [CrossRe
52. Othmen, W.B.H.; Ben Hamed, Z.; Sieber, B.; Addad, A.; Elhouichet, H.; Boukherroub, R. Structural and optical characterizatio of p-type highly Fe-doped SnO2 thin films and tunneling transport on SnO2:Fe/p-Si heterojunction. *Appl. Surf. Sci.* **2018**, *43* 879–890. [CrossRef]
53. Pike, G.E. AC Conductivity of Scandium Oxide and a New Hopping Model for Conductivity. *Phys. Rev. B* **1972**, *6*, 1572–158 [CrossRef]
54. James, A.R.; Prakash, C.; Prasad, G. Structural properties and impedance spectroscopy of excimer laser ablated Zr substitute BaTiO3thin films. *J. Phys. D Appl. Phys.* **2006**, *39*, 1635–1641. [CrossRef]
55. Cole, K.S.; Cole, R.H. Dispersion and Absorption in Dielectrics I. Alternating Current Characteristics. *J. Chem. Phys.* **1941**, 341–352. [CrossRef]
56. Sdiri, N.; Elhouichet, H.; Azeza, B.; Mokhtar, F. Studies of (90-x)P2O5 xB2O3 10Fe2O3 glasses by Mossbauer effect and impedanc spectroscopy methods. *J. Non-Cryst. Solids* **2013**, *371*, 22–27. [CrossRef]
57. Ahmad, I.; Akhtar, M.J.; Younas, M.; Siddique, M.; Hasan, M.M. Small polaronic hole hopping mechanism and Maxwell-Wagne relaxation in NdFeO3. *J. Appl. Phys.* **2012**, *112*, 074105. [CrossRef]
58. Bouslama, W.; Ben Ali, M.; Sdiri, N.; Elhouichet, H. Conduction mechanisms and dielectric constant features of Fe doped ZnC nanocrystals. *Ceram. Int.* **2021**, *47*, 19106–19114. [CrossRef]
59. Awadhia, A.; Patel, S.; Agrawal, S. Dielectric investigations in PVA based gel electrolytes. *Prog. Cryst. Growth Charact. Mate* **2006**, *52*, 61–68. [CrossRef]
60. Raja, V.; Sharma, A.K.; Rao, V.V.R.N. Impedance spectroscopic and dielectric analysis of PMMA-CO-P4VPNO polymer film *Mater. Lett.* **2004**, *58*, 3242–3247. [CrossRef]
61. Langar, A.; Sdiri, N.; Elhouichet, H.; Ferid, M. Structure and electrical characterization of ZnO-Ag phosphate glasses. *Result Phys.* **2017**, *7*, 1022–1029. [CrossRef]
62. Louati, B.; Guidara, K.; Gargouri, M. Dielectric and ac ionic conductivity investigations in the monetite. *J. Alloys. Compd.* **200** *472*, 347–351. [CrossRef]
63. Hammami, H.; Arous, M.; Lagache, M.; Kallel, A. Experimental study of relaxations in unidirectional piezoelectric composite Compos. *Part A Appl. Sci.* **2006**, *37*, 1–8. [CrossRef]
64. Valentin, C.B.S.; Silva, R.L.d.e.; Banerjee, P.; Franco, A. Investigation of Fe-doped room temperature dilute magnetic ZnC semiconductors. *Mat. Sci. Semi. Proc.* **2019**, *96*, 122–126. [CrossRef]
65. Badali, Y.; Altındal, Ş.; Uslu, İ. Dielectric properties, electrical modulus and current transport mechanisms of Au/ZnO/n-S structures. *Prog. Nat. Sci. Mat. Int.* **2018**, *28*, 325–331. [CrossRef]
66. Praveen, K.; Banarji, B.; Srinivas, V.; Choudhary, R.N.P. Complex Impedance Spectroscopic Properties of Ba3V2O8, Ceramics. *Res Lett.* **2008**, *10*, 1–5.
67. Saad, M.; Stambouli, W.; Sdiri, N.; Elhouichet, H. Effect of mixed sodium and vanadium on the electric and dielectric propertie of zinc phosphate glass. *Mat. Res. Bull.* **2017**, *89*, 224–231. [CrossRef]
68. Jlassi, I.; Sdiri, N.; Elhouichet, H. Electrical conductivity and dielectric properties of MgO doped lithium phosphate glasses. *J Non-Cryst. Sol.* **2017**, *466*, 45–51. [CrossRef]

materials

MDPI

Article

Photo-Charging of $Li(Ni_{0.65}Co_{0.15}Mn_{0.20})O_2$ Lithium-Ion Battery Using Silicon Solar Cells

Seungbum Heo [1], Baeksang Yoon [2], Hyunsoo Lim [2], Hyung-Kee Seo [3], Cheul-Ro Lee [2] and Inseok Seo [2,*]

1 Department of Electrical Engineering, Hanbat National University, Daejeon 34158, Korea; heosb3@ex.co.kr
2 School of Advanced Materials Engineering, Jeonbuk National University, Baekje-daero 567, Jeonju 54896, Korea; qortkd2134@jnbu.ac.kr (B.Y.); kady1004@naver.com (H.L.); crlee7@jbnu.ac.kr (C.-R.L.)
3 Future Energy Convergence Core Center, School of Chemical Engineering, Jeonbuk National University, Baekje-daero 567, Jeonju 54896, Korea; hkseo@jbnu.ac.kr
* Correspondence: isseo@jbnu.ac.kr

Abstract: This study reports an integrated device in which a lithium-ion battery (LIB) and Si solar cells are interconnected. The LIB is fabricated using the $Li(Ni_{0.65}Co_{0.15}Mn_{0.20})O_2$ (NCM622) cathode and the $Li_4Ti_5O_{12}$ (LTO) anode. The surface and shape morphologies of the NCM and LTO powders were investigated by field emission scanning electron microscopy (FE-SEM). In addition, the structural properties were thoroughly examined by X-ray diffraction (XRD). Further, their electrochemical characterization was carried out on a potentiostat. The specific discharge capacity of the NCM cathode (half-cell) was 188.09 mAh/g at 0.1 C current density. In further experiments, the NCM-LTO full-cell has also shown an excellent specific capacity of 160 mAh/g at a high current density of 1 C. Additionally, the capacity retention was outstanding, with 99.63% at 1 C after 50 cycles. Moreover, to meet the charging voltage requirements of the NCM-LTO full-cell, six Si solar cells were connected in series. The open-circuit voltage (V_{OC}) and the short-circuit photocurrent density (J_{SC}) for the Si solar cells were 3.37 V and 5.42 mA/cm^2. The calculated fill factor (FF) and efficiency for the Si solar cells were 0.796 and 14.54%, respectively. Lastly, the integrated device has delivered a very high-power conversion-storage efficiency of 7.95%.

Keywords: $Li(Ni_{0.65}Co_{0.15}Mn_{0.20})O_2$; $Li_4Ti_5O_{12}$; lithium-ion battery; silicon solar cell; integrated device

1. Introduction

Environmental problems caused by fossil fuels, such as greenhouse gas (GHG) emissions, drought, and rising sea levels, have recently become major global issues [1,2]. Therefore, many studies have investigated eco-friendly renewable energy sources (solar, wind, geothermal, etc.) that generate electricity with a minimal carbon footprint [3]. Solar energy systems have attracted the most attention among these renewable energy sources as the sun is a popular infinite free energy source. Therefore, photovoltaics (PV) have the potential to meet the global energy demand [4]. However, the intermittent availability of sunlight is a significant drawback that hinders its various applications. In the present study, the idea is to connect the solar cell to LIB directly without any converter. The integration of energy generation and energy storage systems has been presented as one of the better solutions to address the intermittent nature of sunlight [5]. Liao et al. reported the need for a bidirectional DC/DC converter to control a solar power system [6]. The bidirectional DC/DC converter system was more complex and expensive than the direct connection of the solar cell and the energy storage device. Therefore, a direct DC storage system is currently in the spotlight by using the integrated device as a promising technology. Many studies have used capacitors and lithium-ion batteries (LIBs) as storage for charging by PVs [7–11].

LIB is one of the potential energy storage devices used in a wide range of applications, from miniaturized devices to high-power-driven electric vehicles. They are considered suitable energy storage devices that can be charged by solar-powered PVs. It is known that the electrochemical performance of LIB is mainly dependent on the active cathode and anode electrode materials. Among the LIB cathode materials, the $LiFePO_4$ (LFP) cathode material has been widely used in energy storage systems (ESS) because of its low price and high safety. However, LFP has an inferior specific capacity compared to the $Li(Ni_xCo_yMn_z)O_2$ (NCM) cathode material [12–14]. This study proposes to use the NCM622 cathode material. This cathode material is richer in nickel content than NCM333, commonly used in the past, which results in high discharge capacity and energy density.

Moreover, the lower cobalt (Co) content reduces the overall cost of the batteries [15]. On the other hand, $Li_4Ti_5O_{12}$ (LTO) is preferred among the anode materials due to its high-capacity retention rate with a reasonable specific capacity [16]. In addition, LTO also has a stable voltage plateau at 1.5 V vs. Li/Li^+. This high-voltage plateau of LTO can prevent the reduction of electrolytes and can do better than the Li metal anode. Therefore, NCM622 and LTO are chosen to increase the specific capacity and achieve high cycle retention. In this study, as discussed earlier, an integrated device is designed by combining LIB and Si solar cells without any converter. The morphological and structural characterization of the NCM622 cathode powder and the LTO anode powder were evaluated by FE-SEM and XRD, respectively. The electrochemical properties were obtained from the NCM (NCM vs. Li/Li^+) half-cell, LTO (LTO vs. Li/Li^+) half-cell, and NCM-LTO full-cells. Later, six single Si solar cells were connected in series to obtain sufficient voltage to charge the battery. The electrical properties such as open-circuit voltage, short-circuit photocurrent density, fill factor, and efficiency of Si solar cells were analyzed with a solar simulator. Finally, an integrated device was fabricated by directly combining Si solar cells and an NCM-LTO full-cell. Furthermore, the photovoltaic charging of the NCM-LTO full-cell was successfully demonstrated.

2. Materials and Methods

2.1. Preparation of LIB and Solar Cell

The cathode slurry was prepared by dispersing NCM powder (active material), carbon black (super P), and polyvinylidene difluoride (PVDF) in N-Methyl-2-pyrrolidone (NMP) solvent at a respective weight ratio of 8:1:1. Similarly, the anode slurry was prepared using LTO powder as the active material. The mass loading of cathode and anode material was 7.59 and 8.78 mg/cm^2, respectively. Additionally, the densities of the cathode and anode were 3.05 and 2.61 g/cm^3, respectively. The cathode and anode slurries were coated on the aluminum foil and the copper foil, respectively. The coated slurries were then dried and pressed to make sheet electrodes. The half-cells and full-cells were assembled in a glove box filled with argon gas ($H_2O < 0.01$ ppm, $O_2 < 0.01$ ppm), and 1.2 M $LiPF_6$ in a 1:1 (v/v) solution of ethylene carbonate and dimethyl carbonate was used as an electrolyte. Then, 150 μL of electrolyte was added for coin cells. Li metal foil with a thickness of 0.45 mm was used for coin cell fabrication (MTI Korea). The half-cell was fabricated in the form of the coin cell (CR 2032), while the full-cell was assembled in the form of a pouch (4.5 × 4.5 cm). For the pouch cell, nickel and aluminum were used as the cathode and anode lead tabs, and 400 μL of electrolyte was added. The loading and density of electrodes had the same value as the coin cell fabrication. Polyethylene separator was used for every cell fabrication. The electrochemical performance data of half-cells and full-cells corresponds to the coin cell and pouch cell configuration, respectively. A polycrystalline Si solar cell with six single Si solar cells (SAVE SOLAR Co. Ltd. Washington DC, USA) connected in series was integrated with LIB. Therefore, it provides the required voltage for LIB charging.

2.2. Characterizations

The NCM622 precursor was prepared by the co-precipitation method. To improve the packing density of the cathode electrode, the NCM powder was a mixture of large and small particles in a 7:3 ratio. Commercial LTO powder was purchased from Sigma Aldrich Co. Ltd St. Louis, MO, USA. The surface and shape morphologies of the NCM and

LTO powders were characterized by field emission scanning electron microscopy (FE-SEM) (SU-70 Hitachi High-Tech Corporation, Fukuoka, Japan). The acceleration voltage for the FE-SEM measurement was 10 kV. The crystal structures were analyzed using X-ray diffraction (XRD) (MAX-2500, RIGAKU, Tokyo, Japan) with Cu-kα radiation over a 2θ range of 10–80°, with a step size of 0.05° and a scan rate of 2.0°/min. The electrochemical properties of the NCM and LTO half-cells and NCM-LTO full-cell were evaluated using a battery cycler (WBCS3000S, WONATECH Co. Ltd., Seoul, Korea). Galvanostatic charge–discharge tests of the NCM-LTO full-cell were conducted in the 1.6–3.0 V potential window. The NCM-LTO full-cell was tested for rate capability at different current densities. Further, cyclic stability studies were performed on the NCM-LTO full-cell (pouch cell configuration) for 50 cycles at 1 C. The open-circuit voltage (V_{OC}), short-circuit photocurrent density (J_{SC}), fill factor (FF), and efficiency were measured by irradiating Si solar cells with 100 mW/cm^2 light with a solar simulator (PEC-L01, Peccell Technologies, Yokohama, Japan). All tests were performed at room temperature.

2.3. Integrated Si Solar Cells–LIB Device

The integrated device was fabricated by directly connecting the Si solar cells and LIB. The LIB was charged by irradiating light to Si solar cells using a solar simulator (PEC-L01, Peccell Technologies, Yokohama, Japan). Before testing, the light intensity was calibrated to 100 mW/cm^2. The Si solar cells were photo-charged using a solar simulator then discharged with a battery cycler.

3. Results and Discussion

Figure 1 displays the FE-SEM images of the NCM and LTO powders. The particle size and shape of the NCM powder are shown in Figure 1a,b. Figure 1a shows that NCM powder was composed of large and small particles. The NCM powder consists of large and small particles with respective sizes of about 10 and 3 μm; additionally, the large and small particles were mixed in an optimized 7:3 weight ratio, respectively. The optimization was performed with an objective to improve the high-energy density [17]. Figure 1b shows the agglomerated particles of NCM powder. Submicron particles were aggregated to form large and small particles. Figure 1c,d illustrate the particle morphology of LTO powder. Figure 1c shows that the LTO powder contains porous spherical particles with a diameter of about 20 μm. The spherical shape improves the high rate performance of the LTO by minimizing the Li-ion diffusion path lengths [18]. Figure 1d shows a detailed view of the porous spherical particles of LTO powder. The porous spherical particles are made from aggregated particles with a size of less than about 0.5 μm.

Figure 2a,b, respectively, depict the XRD patterns of the NCM and LTO powders. As shown in Figure 2, the XRD patterns of NCM and LTO powder were identified and indexed with the powder diffraction database (PDF). The XRD peaks of NCM powders around 18.7°, 36.6°, 37.9°, 38.7°, 44.4°, 48.5°, 58.5°, 64.3°, 64.8°, and 68.1° are assigned to the (0 0 3), (1 0 1), (0 0 6), (0 1 2), (1 0 4), (0 1 5), (1 0 7), (0 1 8), (1 1 0), and (1 1 3) planes, respectively. The diffraction peaks of LTO powders around 18.3°, 35.5°, 37.1°, 43.2°, 47.3°, 57.1°, 62.7°, 66.0°, 74.2°, 75.2°, and 79.2° are assigned to the (1 1 1), (3 3 1), (2 2 2), (4 0 0), (3 3 1), (3 3 3), (4 4 0), (5 3 1), (5 3 3), (6 2 2), and (4 4 4) planes, respectively. The XRD measurement results showed that the NCM and LTO powders were successfully synthesized.

Figure 3a–c show the electrochemical properties at 0.1, 0.2, 0.5, and 1 C of the NCM-Li, LTO-Li half-cells, and the NCM-LTO full-cell, respectively. Figure 3a reveals the discharge capacities of 188.09, 182.92, 173.27, and 162.81 mAhg^{-1} at 0.1, 0.2, 0.5, and 1 C, respectively. The discharge capacities of the NCM-Li half-cell were similar to those of reported values in the previous study [19]. Figure 3b shows that the discharge capacities of LTO-Li half-cells were 161.3, 159.76, 159.7, and 157.57 mAhg^{-1} at 0.1, 0.2, 0.5, and 1 C, respectively. Further, the NCM-LTO full-cell has delivered a stable specific discharge capacity of 185.79 mAhg^{-1} at 0.1 C, 182.02 mAhg^{-1} at 0.2 C, 173.19 mAhg^{-1} at 0.5 C, and 160.48 mAhg^{-1} at 1 C. The discharge capacity of the NCM-LTO full-cell was very similar to that of the NCM-Li half-cell. Additionally, the operating voltage of the full-cell was near about 2.3 V. This operating voltage is reasonable for integrating solar cells with the battery. Figure 3d shows the discharge capacities of the NCM-LTO full-cell at various C-rates. The NCM-LTO full-cell showed the same discharge capacities during five cycles at the same C-rate, which supports the stable performance of the NCM-LTO full-cell.

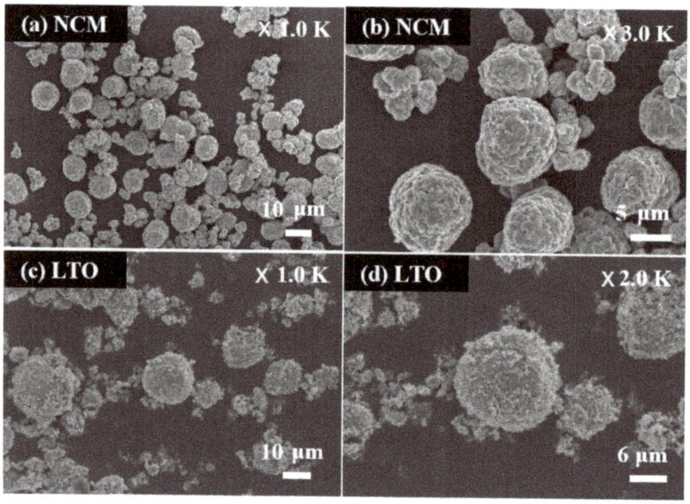

Figure 1. FE-SEM images of (**a**,**b**) NCM powder and (**c**,**d**) LTO powder.

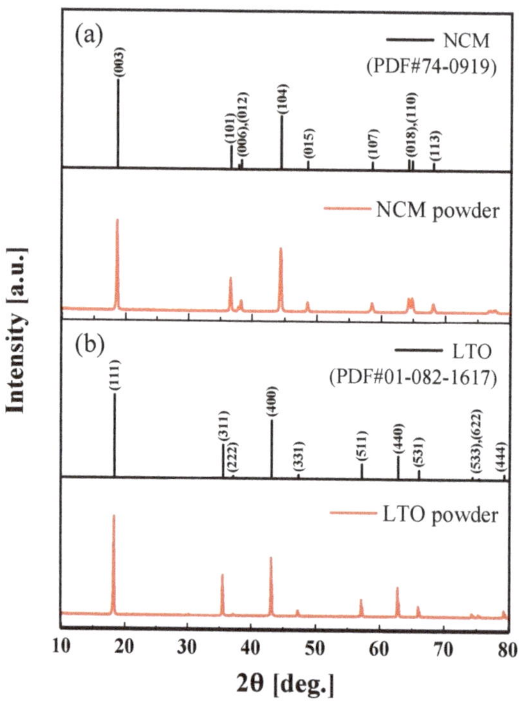

Figure 2. XRD patterns of (**a**) NCM (PDF#74-0919) powder and (**b**) LTO (PDF#01-082-1617) powder.

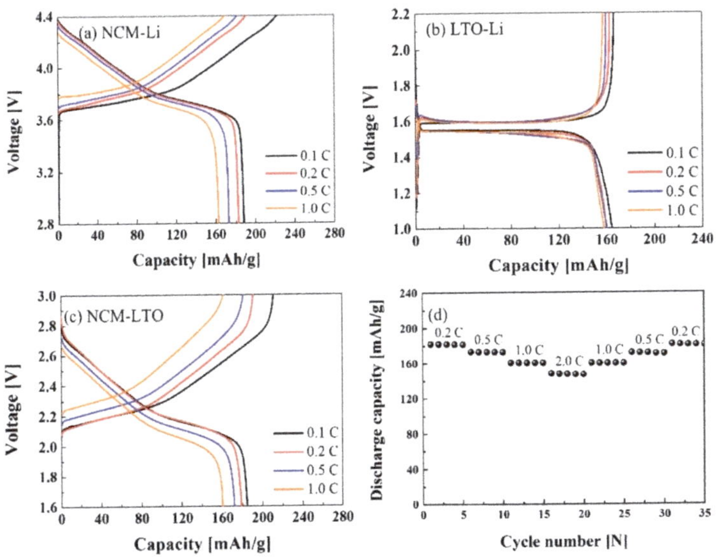

Figure 3. Electrochemical properties of (**a**) NCM-Li, (**b**) LTO-Li, and (**c**) NCM-LTO full-cells, and (**d**) rate capability of the NCM-LTO full-cell at various C-rates.

Figure 4a,b show the results of cycle tests for 50 cycles of the NCM-LTO full-cell at 1 C-rate. The charge–discharge was carried out in the 1.6–3.0 V range. The discharge capacities were 160.48 $mAhg^{-1}$ at the 1st cycle, 160.32 $mAhg^{-1}$ at the 5th cycle, 159.97 $mAhg^{-1}$ at the 25th cycle, and 159.89 $mAhg^{-1}$ at the 50th cycle. The NCM-LTO full-cell exhibited excellent cycle retention of 99.63% even after 50 cycles. Figure 4b shows the discharge capacities and coulombic efficiency with respect to the cycle number for 50 cycles. In cycles 1, 5, 25, and 50, the coulombic efficiencies were 99.99%, 99.98%, 99.97%, and 99.95%, respectively. This shows the exceptional stability of the electrode materials.

Figure 5 shows a schematic diagram of the integrated device using the Si solar cell and the NCM-LTO full-cell. The voltage range of the NCM-LTO full-cell was 1.6–3.0 V. The necessary charging voltage was obtained by connecting six polycrystalline single Si solar cells of 0.57 V each in series. The integrated device was fabricated by connecting directly without a converter. Using this system, the integrated device can charge the battery through photo-charging.

Figure 6 depicts the current density-voltage (J-V) characteristics of the interconnected Si solar cells. The short-circuit photocurrent density (J_{SC}) and open-circuit voltage (V_{OC}) of a Si solar cell were 5.42 mA/cm^2 and 3.37 V. The interconnected Si solar cells can provide the required voltage to charge the NCM-LTO full-cell. The fill factor (FF) and efficiency of the Si solar cells were calculated to be 0.796 and 14.54%, respectively. The Si solar cells used in this study have displayed a slightly higher efficiency of 14.54%, compared to commercially available Si solar cells, which are typically about 12–15% [20].

Figure 7 shows the electrochemical results of the integrated device. The charge–discharge tests were carried out for 20 cycles in the 1.6–3.0 V range. The integrated device was charged by the Si solar cells using a solar simulator and discharged by a battery cycler to the 10th cycle. After that, the NCM-LTO full-cell was charged and discharged for 11 to 20 cycles using a battery cycler. The current density of the integrated device was 40 mA. The photo-charging and galvanostatic discharging profiles from 1 to 10 cycles were similar to those of galvanostatic charge and discharge (using battery cycler) from 11 to 20 cycles. The power conversion/storage efficiency of the integrated device was 7.95%. Moreover, energy storage efficiency from the Si solar cells to the NCM-LTO full-cell was 54.36%. Low-power conversion efficiency has represented an essential obstacle in researching integrated systems wherein Si solar cells and batteries are connected. It is expected that the integrated device fabricated in this study can figure out the issues of low-power conversion efficiency [10].

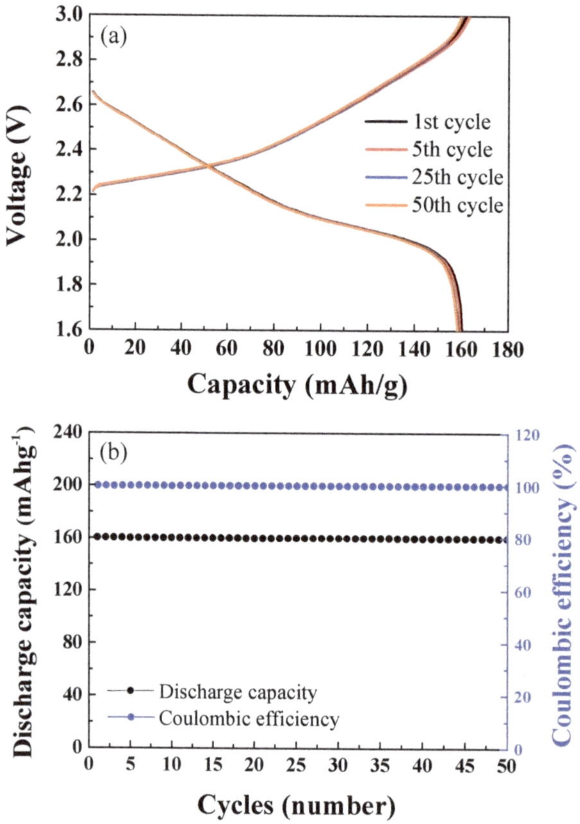

Figure 4. (**a**) Cycle performance of the NCM-LTO full-cell over 50 cycles. (**b**) Discharge capacity and coulombic efficiency of the NCM-LTO full-cell over 50 cycles.

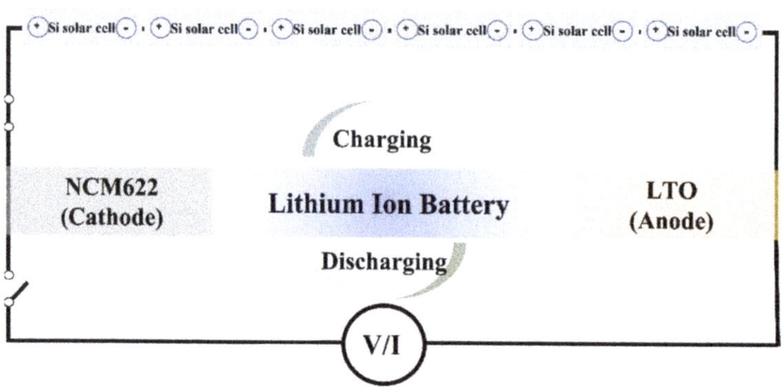

Figure 5. Schematic diagram of the NCM-LTO integrated device with Si solar cells.

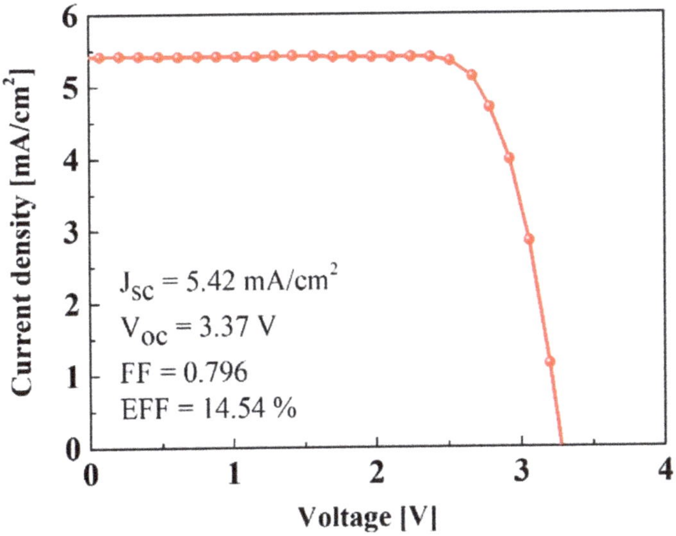

Figure 6. J-V curve of the Si solar cells.

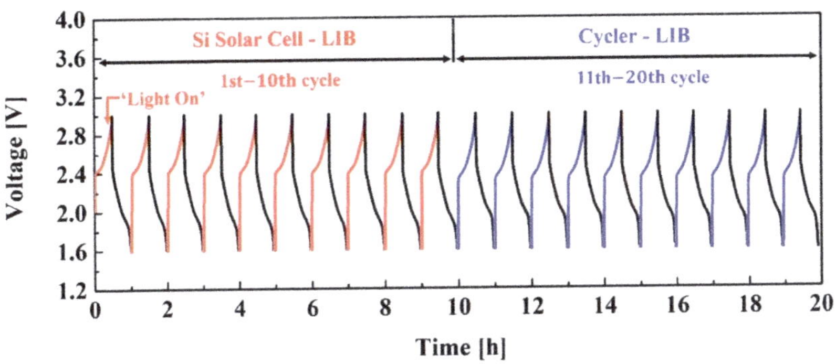

Figure 7. Photo-charging and discharging diagram of the NCM-LTO full-cell using the Si solar cell.

Figure 8a shows the J-V characteristics of the Si solar cells connected to the NCM-LTO full-cell before cycle testing and after 1st, 5th, and 10th cycle tests. The open-circuit voltage (V_{OC}) was 3.37 V before the cycle test, and 3.37, 3.36, and 3.35 V during the 1st, 5th, and 10th cycles, respectively. The short-circuit photocurrent density (J_{SC}) has remained constant with increasing cycle numbers. The J_{SC} was 5.42 mA/cm^2 before the cycle test, and 5.42, 5.41, and 5.4 mA/cm^2, during the 1st, 5th, and 10th cycles, respectively. Figure 8b shows the energy efficiency and fill factor (FF) of the Si solar cells connected to the NCM-LTO full-cell before and after various cycle tests. Before cycle testing, the fill factor (FF) was 0.796. Then, the FF was 0.793, 0.793, and 0.792 at the 1st, 5th, and 10th cycles, respectively. These results show that the FFs were almost unchanged during the test. The efficiency was 14.54% before the cycle test, and after the 1st, 5th, and 10th cycle tests, it slightly reduced to 14.48%, 14.42%, and 14.35%, respectively. The energy conversion efficiency of the Si solar cells connected to the NCM-LTO full-cell was 98.69% before and after 10 cycles. Therefore, the Si solar cells were suitable as a device for supplying power to the high-performance NCM-LTO full-cell and can pave the way to futuristic energy-saving technologies.

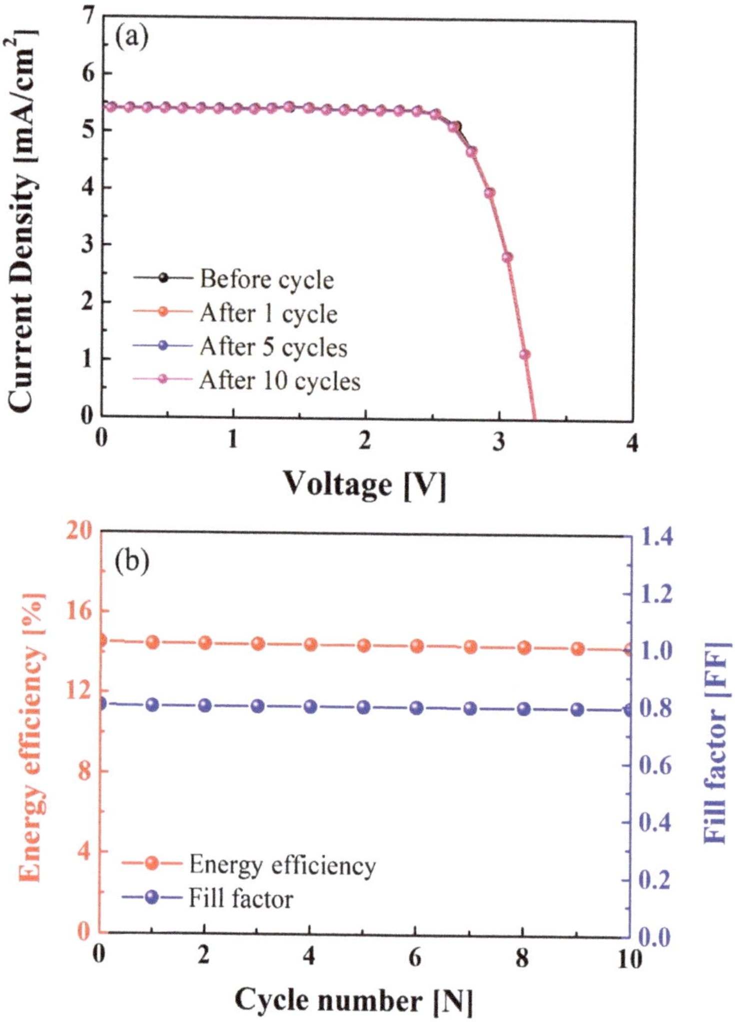

Figure 8. (**a**) J-V curve and (**b**) energy efficiency and fill factor (FF) of Si solar cells connected to the NCM-LTO full-cell before and after 10 cycles.

4. Conclusions

The NCM-LTO full-cell was fabricated using NCM and LTO as the cathode and the anode, respectively. The as-fabricated full-cell has delivered a high discharge capacity and stable performance, with 99.63% cycle retention after 50 cycles. This high performance is mainly due to the synergy of the optimized NCM622 cathode and LTO anode materials compared to conventional LFP-graphite batteries. Besides, the NCM-LTO full-cell is directly interconnected with six Si solar cells without any convertor. The integrated device was photo-charged by the Si solar cell and galvanostatic discharged by a cycler. The power conversion efficiency of the NCM-LTO full-cell and Si solar cells was 7.95%, while the energy conversion efficiency reached 54.36%. Therefore, the present integrated system using the NCM-LTO full-cell could be a promising technology for the next-generation self-charging devices.

Author Contributions: Writing—original draft, S.H.; formal analysis, B.Y.; data curation, H.L.; methodology, H.-K.S.; conceptualization, C.-R.L.; formal analysis, funding acquisition and supervision, I.S. All authors have read and agreed to the published version of the manuscript.

Funding: This work was supported by a National Research Foundation of Korea (NRF) grant funded by the Korean government (MSIT) (NRF-2020R1A2C2009057). This work was also partially supported by a Korea Basic Science Institute (National Research Facilities and Equipment Center) grant funded by the Ministry of Education (grant 2020R1A6C101A204).

Institutional Review Board Statement: Not applicable.

Informed Consent Statement: Not applicable.

Data Availability Statement: The data presented in this study are available on request from the corresponding author. The data are not publicly available as it is part of the ongoing project.

Conflicts of Interest: The authors declare no conflict of interest.

References

1. Lewis, N.S. Research Opportunities to Advance Solar Energy Utilization. *Science* **2016**, *351*, aad1920. [CrossRef] [PubMed]
2. Hosseini, S.E.; Wahid, M.A.; Aghili, N. The Scenario of Greenhouse Gases Reduction in Malaysia. *Renew. Sustain. Energy Rev.* **2013**, *28*, 400–409. [CrossRef]
3. Rabaia, M.K.H.; Abdelkareem, M.A.; Sayed, E.T.; Elsaid, K.; Chae, K.-J.; Wilberforce, T.; Olabi, A.G. Environmental Impacts of Solar Energy Systems: A Review. *Sci. Total Environ.* **2021**, *754*, 141989. [CrossRef] [PubMed]
4. Blaschke, T.; Biberacher, M.; Gadocha, S.; Schardinger, I. 'Energy Landscapes': Meeting Energy Demands and Human Aspirations. *Biomass Bioenergy* **2013**, *55*, 3–16. [CrossRef] [PubMed]
5. Um, H.-D.; Choi, K.-H.; Hwang, I.; Kim, S.-H.; Seo, K.; Lee, S.-Y. Monolithically Integrated, Photo-Rechargeable Portable Power Sources Based on Miniaturized Si Solar Cells and Printed Solid-State Lithium-Ion Batteries. *Energy Environ. Sci.* **2017**, *10*, 931–940. [CrossRef]
6. Liao, Z.; Ruan, X. Control Strategy of Bi-Directional DC/DC Converter for a Novel Stand-Alone Photovoltaic Power System. In Proceedings of the 2008 IEEE Vehicle Power and Propulsion Conference, Harbin, China, 3–5 September 2008; Volume 23.
7. Hauch, A.; Georg, A.; Krašovec, U.O.; Orel, B. Photovoltaically Self-Charging Battery. *J. Electrochem. Soc.* **2002**, *149*, A1208. [CrossRef]
8. Miyasaka, T.; Murakami, T.N. The Photocapacitor: An Efficient Self-Charging Capacitor for Direct Storage of Solar Energy. *Appl. Phys. Lett.* **2004**, *85*, 3932–3934. [CrossRef]
9. Murakami, T.N.; Kawashima, N.; Miyasaka, T. A High-Voltage Dye-Sensitized Photocapacitor of a Three-Electrode System. *Chem. Commun.* **2005**, *26*, 3346–3348. [CrossRef] [PubMed]
10. Guo, W.; Xue, X.; Wang, S.; Lin, C.; Wang, Z.L. An Integrated Power Pack of Dye-Sensitized Solar Cell and Li Battery Based on Double-Sided TiO_2 Nanotube Arrays. *Nano Lett.* **2012**, *12*, 2520–2523. [CrossRef] [PubMed]
11. Yang, Z.; Li, L.; Luo, Y.; He, R.; Qiu, L.; Lin, H.; Peng, H. An Integrated Device for Both Photoelectric Conversion and Energy Storage Based on Free-Standing and Aligned Carbon Nanotube Film. *J. Mater. Chem. A* **2013**, *1*, 954–958. [CrossRef]
12. Nitta, N.; Wu, F.; Lee, J.T.; Yushin, G. Li-Ion Battery Materials: Present and Future. *Mater. Today* **2015**, *18*, 252–264. [CrossRef]
13. Padhi, A.K.; Nanjundaswamy, K.S.; Goodenough, J.B. Phospho-olivines as Positive-Electrode Materials for Rechargeable Lithium Batteries. *J. Electrochem. Soc.* **1997**, *144*, 1188–1194. [CrossRef]
14. Jiang, L.; Wang, Q.; Sun, J. Electrochemical Performance and Thermal Stability Analysis of LiNi(x)Co(y)Mn(z)O(2) Cathode Based on a Composite Safety Electrolyte. *J. Hazard. Mater.* **2018**, *351*, 260–269. [CrossRef] [PubMed]
15. Choi, J.U.; Voronina, N.; Sun, Y.-K.; Myung, S.-T. Recent Progress and Perspective of Advanced High-Energy Co-Less Ni-Rich Cathodes for Li-Ion Batteries: Yesterday, Today, and Tomorrow. *Adv. Energy Mater.* **2020**, *10*, 2002027. [CrossRef]
16. Yi, T.-F.; Jiang, L.-J.; Shu, J.; Yue, C.-B.; Zhu, R.-S.; Qiao, H.-B. Recent Development and Application of $Li_4Ti_5O_{12}$ as Anode Material of Lithium Ion Battery. *J. Phys. Chem. Solids* **2010**, *71*, 1236–1242. [CrossRef]
17. Park, C.; Lee, S.; Kim, K.; Kim, M.; Choi, S.; Shin, D. Electrochemical Properties of Composite Cathode Using Bimodal Sized Electrolyte for All-Solid-State Batteries. *J. Electrochem. Soc.* **2019**, *166*, A5318–A5322. [CrossRef]
18. Shen, L.; Yuan, C.; Luo, H.; Zhang, X.; Xu, K.; Xia, Y. Facile Synthesis of Hierarchically Porous $Li_4Ti_5O_{12}$ Microspheres for High Rate Lithium Ion Batteries. *J. Mater. Chem.* **2010**, *20*, 6998–7004. [CrossRef]
19. Noh, H.-J.; Youn, S.; Yoon, C.S.; Sun, Y. Comparison of the Structural and Electrochemical Properties of Layered $Li[Ni_xCo_yMn_z]O_2$ (x = 1/3, 0.5, 0.6, 0.7, 0.8 and 0.85) Cathode Material for Lithium-Ion Batteries. *J. Power Sources* **2013**, *233*, 121–130. [CrossRef]
20. Sharma, S.; Jain, K.K.; Sharma, A.; Sharma, S.; Jain, K.K.; Sharma, A. Solar Cells: In Research and Applications—A Review. *Mater. Sci. Appl.* **2015**, *6*, 1145–1155. [CrossRef]

materials

MDPI

Article

Material Optimization Engineering toward $xLiFePO_4 \cdot yLi_3V_2(PO_4)_3$ Composites in Application-Oriented Li-Ion Batteries

Yuqiang Pi [1], Gangwei Luo [1], Peiyao Wang [2], Wangwang Xu [3], Jiage Yu [1], Xian Zhang [1], Zhengbing Fu [1], Xiong Yang [1], Li Wang [1], Yu Ding [1,2,4,*] and Feng Wang [1,*]

[1] School of Chemistry and Materials Science, Hubei Engineering University, Xiaogan 432000, China; xq0502@163.com (Y.P.); 15885400892@163.com (G.L.); yujiage5@163.com (J.Y.); zxian821@163.com (X.Z.); ceramic425@163.com (Z.F.); yangx@hbeu.edu.cn (X.Y.); wangli19810609@126.com (L.W.)
[2] Department of Mechanical Engineering, The University of Melbourne, Parkville, VIC 3010, Australia; peiyao.wang@student.unimelb.edu.au
[3] College of Materials and Chemical Engineering, China Three Gorges University, Yichang 443002, China; wxu26@lsu.edu
[4] Qinghai Electronic Material Industry Development Co., Ltd., Xining 810006, China
* Correspondence: dy9802@126.com (Y.D.); hbeukj@126.com (F.W.)

Abstract: The development of $LiFePO_4$ (LFP) in high-power energy storage devices is hampered by its slow Li-ion diffusion kinetics. Constructing the composite electrode materials with vanadium substitution is a scientific endeavor to boost LFP's power capacity. Herein, a series of $xLiFePO_4 \cdot yLi_3V_2(PO_4)_3$ (xLFP·yLVP) composites were fabricated using a simple spray-drying approach. We propose that 5LFP·LVP is the optimal choice for Li-ion battery promotion, owning to its excellent Li-ion storage capacity (material energy density of 413.6 $W \cdot h \cdot kg^{-1}$), strong machining capability (compacted density of 1.82 $g \cdot cm^{-3}$) and lower raw material cost consumption. Furthermore, the 5LFP·LVP| |LTO Li-ion pouch cell also presents prominent energy storage capability. After 300 cycles of a constant current test at 400 mA, 75% of the initial capacity (379.1 mA·h) is achieved, with around 100% of Coulombic efficiency. A capacity retention of 60.3% is displayed for the 300th cycle when discharging at 1200 mA, with the capacity fading by 0.15% per cycle. This prototype provides a valid and scientific attempt to accelerate the development of xLFP·yLVP composites in application-oriented Li-ion batteries.

Keywords: $xLiFePO_4 \cdot yLi_3V_2(PO_4)_3$; spray-drying; power capacity; compacted density; Li-ion batteries

1. Introduction

The commitment to carbon neutrality and the implementation of a low-carbon economy in the world have become the key to the sustainable development of human society, given the massive population explosion and incremental demand for new or renovated infrastructure [1,2]. The most reliable technique for reducing carbon emissions is to increase the utilization of available renewable energy sources [3–5]. The energy storage technologies (ESTs) are critical to attaining renewable energy sources with intermittent and unstable characteristics. Among the variety of ESTs, lithium-ion batteries (LIBs) have been widely recognized in the daily life of human beings, due to their high-energy density and ultralong lifespan [6–8].

Cathode material as a key component determines many aspects of LIBs, including the Li-ion storage capability, lifespan, cost, and safety [9,10]. In particular, olivine $LiFePO_4$ (LFP) has received substantial attention from academia and industry, because of its relatively high theoretical energy density (580 $W \cdot h \cdot kg^{-1}$), structural integrity during the Li-ion extraction/insertion process, and low raw cost consumption [11,12]. At present, LFP research is primarily focused on three aspects: (i) increasing discharge specific capacity,

(ii) enhancing the power capability, and (iii) improving the low-temperature performance. In particular, the sluggish Li-ion diffusion kinetics of LFP impedes its development in high-power energy storage devices [13,14]. Many attempts have been made to favor Li^+ fast migration in order to address this issue. One of them is to modify LFP by combining it with $Li_3V_2(PO_4)_3$ (LVP), which has three-dimensional pathways for Li^+ mobility [15,16]. However, such a strategy proposed by many researchers mainly aims at improving electrochemical performance, leaving out application-oriented research on the compacted density and Li-ion pouch cells [17–20].

In addition to selecting a cathode material with a high Li-ion diffusion coefficient, pursuing high-power LIBs is still dependent on several factors. Firstly, the anode material should be carefully considered. $Li_4Ti_5O_{12}$ (LTO) is an appealing anode candidate for high-power LIBs, because of its fast Li^+ transport capability, zero volume change during charging cycles, and alleviated lithium deposition [21,22]. Secondly, the component ratio of electrode pieces can be fine-tuned to adapt to rapid Li-ion intercalation and deintercalation behavior. Moreover, the use of high-conductivity additive, such as carbon nanotubes, can significantly improve the electron transfer capability of electrode pieces. Thirdly, another strategy is to replace traditional foils with new high-performance foils. For example, carbon-coated Al foils enable lower internal resistance, decreased polarization, and improved adhesion between active materials and the current collector. They provide an opportunity for LIBs to achieve a higher power density with a considerable energy density. In practice, the compacted density is a critical parameter of battery design, and the volumetric energy density of the battery can be indicated directly by the compacted density of electrode pieces, which is determined by the size distribution of active material [23,24], the components of electrode pieces, and the imposed pressure from roller machine [25].

To develop an application-oriented energy storage device with a high power density, material engineering optimization was conducted to manufacture a series of xLFP·yLVP composites including LFP·LVP, 3LFP·LVP, 5LFP·LVP, and 8LFP·LVP utilizing a simple spray-drying approach. 5LFP·LVP (5LFVP) stands out as a strong candidate cathode material for the high-power LIBs, due to its comprehensive superiorities of Li-ion storage capability, machining capability, and raw material cost consumption. Subsequently, a 5LFVP| |LTO Li-ion pouch cell was exploited to demonstrate its significant energy storage capability. According to the characterization of Li-ion extraction/insertion capability, this prototype Li-ion full cell presents a reasonable and scientific attempt to promote the development of xLFP·yLVP composites in application-oriented Li-ion batteries.

2. Materials and Methods

The information on the crystallographic structure of synthetic products was obtained using a D8 Advance X-ray diffractometer (XRD, Bruker AXS, Karlsruhe, Germany) with a non-monochromatic Cu K α X-ray source. The features of amorphous carbon in xLFP·yLVP composites were determined using the Renishaw INVIA micro-Raman spectroscopy system (Gloucestershire, UK). The Vario EL cube CHNSO elemental analyzer (Hanau, Germany) was employed to examine the carbon content. The information of the valence state was evaluated by XPS measurement via an Ultra DLD with a monochromic Al X-ray source. The Microtrac S3500 laser (Microtrac Inc., Largo, FL, USA) particle size analyzer was used for the particle distribution. To determine the morphology of xLFP·yLVP, the JEOL-7100F (JEOL Ltd., Akishima, Tokyo, Japan) with energy-dispersive spectroscopy (Oxford EDS IE25, Oxford, UK) was employed. The detailed structural information was obtained using a JEM-2100F microscope (JEOL Ltd., Akishima, Tokyo, Japan).

3. Electrochemical Characterization

Li-ion half cells were assembled for estimating the difference in Li-ion storage capabilities between various xLFP·yLVP composites. In this study, the Li-ion half cells were divided into five components: Li foil, power-type electrolyte, separator films from Celgard company, 2025 coin-type battery shell, and electrode piece with 14 mm diameter.

Each electrode piece had a weight ratio of 80:10:10, corresponding to xLFP·yLVP active materials, conductive carbon nanotube paste (5 wt.%, Shandong Chenghe New Material Co., Ltd., Zibo, China) and PVDF binder, respectively. The current collector was made of carbon-coated aluminum foil. The mass loading of all xLFP·yLVP electrodes was controlled between 2.0 and 3.0 mg·cm^{-2}. The LAND CT2001A battery test systems were used to characterize the electrochemical performance. To evaluate the electrochemical behavior, cyclic voltammetry (CV) curves were collected using a CHI 760e electrochemical workstation. To achieve a reasonable pouch-cell capacity design, the half cells of LTO| |Li were also assembled using the same method described above to assess the specific capacity and cyclability of anode material.

The pouch-cell fabrication allows a rational way to demonstrate the utilization of xLFP·yLVP composites in application-oriented Li-ion batteries. To meet the requirements of practical application, the ratio of electrode materials, for all types of xLFP·yLVP cathodes and LTO anode, was set to 90.5 wt.%. The contents of carbon nanotube conductive paste, super P, and PVDF binder in the pouch cells were adjusted subsequently to 1.5 wt.%, 2 wt.%, and 6 wt.%, respectively. Rolls of separator films were used to prevent direct contact between the cathode and anode pieces. Here, we proceeded with the control on a capacity ratio of the anode to cathode, which could fluctuate within the range 1–1.05. Moreover, the roller machine with 15 tons of pressure was performed to achieve the appropriate compacted densities of all types of xLFP·yLVP cathodes and LTO anode pieces. A battery test system of TITANS THCX-IO/5-96-M-A was used to estimate the electrochemical property of pouch cells.

4. Synthetic Procedure

In this work, material engineering optimization was conducted to manufacture the xLFP·yLVP composites using a simple spray-drying approach. Here, the molar ratio of x/y was varied from 1:1, 3:1, 5:1, to 8:1 with corresponding products denoted as LFVP, 3LFVP, 5LFVP, and 8LFVP, respectively. Scheme 1 depicts the detailed synthesis procedure of 3LFVP, which was chosen as a representative.

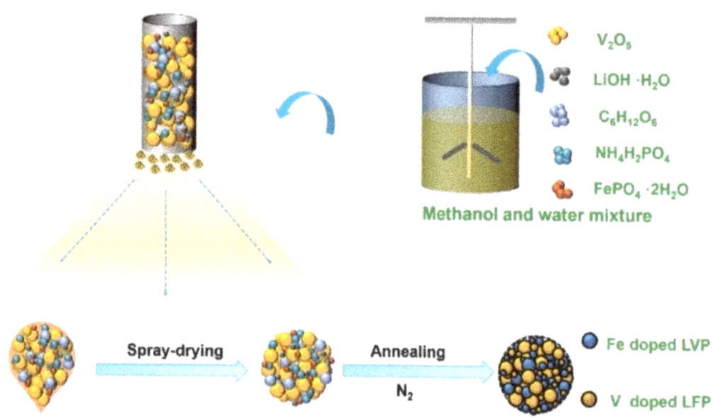

Scheme 1. Schematic representation for manufacturing xLFP·yLVP composites.

A sand-milling machine was operated to make all solid raw materials, i.e., 0.328 kg of V_2O_5, 1.009 kg of $FePO_4·2H_2O$, 0.465 kg of $LiOH·H_2O$, 0.621 kg of $NH_4H_2PO_4$, 0.5 kg of $C_6H_{12}O_6$, and the milling solvent (10 L of methanol and 5 L of water) was mixed to form a uniform precursor liquid. The milling time was set at 2 h. The resulting slurry was then transformed into the precursor powder via the continuous heating exchange from the spray-drying equipment. Finally, the abovementioned precursors were sintered in N_2 atmosphere at 700 °C for 12 h, allowing consistent xLFP·yLVP composites. Other xLFP·yLVP composites could be obtained by adjusting the ratio of V_2O_5 to $FePO_4·2H_2O$ in the same synthesized procedure. Moreover, according to the theoretical content calculation of LFP and LVP in xLFP·yLVP composites (Table S1), the components could be recognized.

5. Results and Discussion

Figure 1a shows the XRD patterns of LFVP, 3LFVP, 5LFVP, and 8LFVP. As presented in Figure S1, the diffraction peaks were identified for orthorhombic $LiFePO_4$ (JCPDS No. 81-1173) and monoclinic $Li_3V_2(PO_4)_3$ (JCPDS No. 01-072-7074) [26–29]. There were no undesired impurities in this series of composites, indicating the great potential of our adopted material optimization approach in practical application. When comparing all Raman curves of xLFP·yLVP composites in Figure 1b, two apparent peaks near 1350 cm^{-1} and 1590 cm^{-1} could be indexed to D and G bands, respectively, indicating graphite defects and stretching vibration from sp^2 carbon atom bonds [17,30]. However, it should be noted that the graphitization degree of composites showed comparatively subtle differences with increasing LFP content. The I_D/I_G ratio for LFVP was only 1.02, lower than that of 8LFVP (1.15), suggesting that the amorphous carbon embedded on the LVP surface was inclined to the higher electron conductivity, or that vanadium may have had a positive impact on the formation of graphitized carbon. Furthermore, XPS measurement was carried out to obtain the valence state information of key elements in xLFP·yLVP composites. All peaks labeled in Figure 1c represent the index signals for Li 1*s*, P 2*p*, P 2*s*, C 1*s*, V 2*p*, O 1*s*, and Fe 2*p*. Figure 1d presents the XPS curve of Fe 2*p* with two peaks appearing at around 709 eV and 723 eV for Fe $2p_{3/2}$ and Fe $2p_{1/2}$, respectively [31,32]. In particular, the intensity of Fe 2*p* peaks increased gradually with the increase in the raw material proportion of $FePO_4{\cdot}2H_2O$.

The SEM picture of as-prepared xLFP·yLVP samples in Figures 1e and S2 shows a uniform microsphere structure constructed by our spray-drying technique. The EDS analysis was carried out on the representative selection of 8LFVP, indicating a homogeneous distribution of V, Fe, and C elements in the composites. The TEM and HRTEM measurements provide more detailed information on the structure of 8LFVP, revealing a carbon coating layer outside particles (Figure 1g). Carbon coating has been well documented as an effective modification method for electrode materials with low intrinsic electronic conductivity, allowing for reduced grain growth, enhanced electronic transmission rate, reduced side reaction, and retained structure integrality [14,33–35]. Additionally, lattice spacings of 0.542 nm and 0.421 nm were measured, as marked in Figure 1h,i, assigned to the (111) plane of LVP and (101) plane of LFP, respectively.

To determine the differences in intrinsic kinetics, different scan rates of CV measurements were carried out on the series of xLFP·yLVP composites, and the curves are shown in Figure 2a,c,e,g with the potential range of 2.5–4.3 V, corresponding to LFVP, 3LFVP, 5LFVP, and 8LFVP samples, respectively. The redox peak labeled as a/a′ was ascribed to Li-ion reversible extraction/insertion from/into the LFP crystal structure [36,37]. According to previous reports [38,39], the redox reactions for V^{3+}/V^{4+} from the LVP electrode can induce three pairs of peak signals, which are shown in the figures except for a/a′. The positions of redox peaks are shifted gradually, associated with enlarged electrochemical polarization, when the scan rates are increased. Furthermore, it is well documented that the electrochemical kinetics is determined by the Li-ion diffusion coefficients, which can be estimated from CV results. Here, the a/a′ and b/b′ redox peaks were chosen to calculate the Li-ion diffusion coefficients (D_{Li^+}) of LFP and LVP, respectively. Figure 2b,d,f,h present the linear fittings of I_p vs. $v^{1/2}$, rooting from the Randles–Sevchik equation [34,40]. By calculating the slope of the fitting line, the D_{Li^+} values of xLFP·yLVP composites were obtained and plotted as a contrast diagram (Figure 2i,j). The D_{Li^+} of LFP in 3LFVP was $2.94 \times 10^{-10}/3.1 \times 10^{-10}$ $cm^2{\cdot}S^{-1}$, faster than that of other xLFP·yLVP composites. Meanwhile, the D_{Li^+} of LVP presented similar trend, with 3LFVP having the highest D_{Li^+} value of $8.27 \times 10^{-9}/7.2 \times 10^{-9}$ $cm^2{\cdot}S^{-1}$. Therefore, it is supposed that 3LFVP could provide better reversibility and the fastest Li-ion migration. However, we should note that the Li-ion diffusion rate in LVP exceeded that in LFP by an order of magnitude, but LFP would deliver higher intrinsic specific capacity. The optimal xLFP·yLVP composition would depend on the specific application requirements.

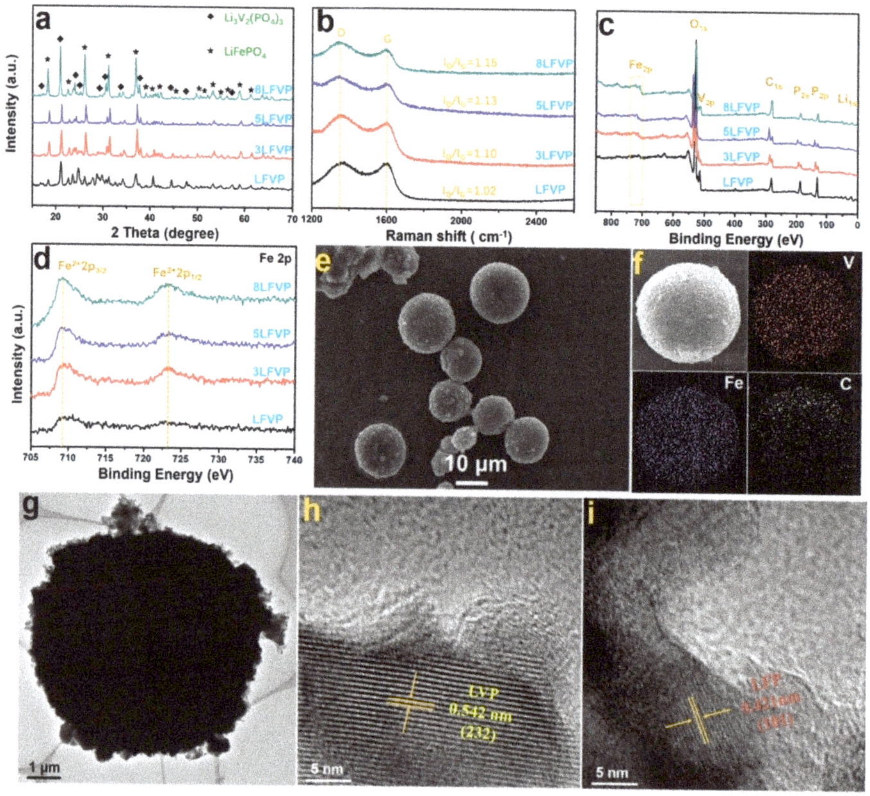

Figure 1. (**a–c**) XRD patterns, Raman spectra, and full XPS spectra of the series of xLFP·yLVP composites; (**d**) Fe 2*p* fine spectrum of xLFP·yLVP composites; (**e**,**f**) SEM images of 8LFVP/C with EDS mapping; (**g**) TEM of 8LFVP; (**h**,**i**) HRTEM of 8LFVP with the lattice spacings of LVP and LFP.

Figure 3a shows the power capabilities of the series of xLFP·yLVP composites, which were evaluated at the step current rate from 0.5 C to 100 C. We can find that, even at the ultrahigh current rate of 100 C, the electrodes could deliver the capacities of 93.3, 70.8, 67, and 44.3 mA·h·g^{-1} for LFVP, 3LFVP, 5LFVP, and 8LFVP electrodes, respectively. Higher specific capacity could be achieved for the 8LFVP electrode at currents less than 2 C. However, when increasing the current rate to 5 C or more, LFVP electrode took the lead due to its high proportion of LVP with a higher ion diffusion coefficient. Figure 3b shows the galvanostatic charge–discharge curves of series of xLFP·yLVP at 0.2 C. It is worth noting that four couples of charge and discharge platforms could be observed for each xLFP·yLVP, agreeing with the results of CV analysis. Moreover, as shown in Figure 3c, when a constant current of 1 C was imposed on each xLFP·yLVP, the initial capacities of 112.7, 111.7, 118.6, and 119.2 mA·h·g^{-1} could be obtained for LFVP, 3LFVP, 5LFVP, and 8LFVP electrodes, respectively. Furthermore, 95.7%, 100%, 97.2%, and 96% capacity retention can be maintained after 1000 cycles, demonstrating excellent structural stability. Although the capacity difference between xLFP·yLVP composites was very small at 1 C, the material energy density varied significantly, which was closely related to discharge medium voltage. As shown in Figure 3d, 5LFVP stood out from the other three electrodes, achieving the highest material energy density of 413.6 W·h·kg^{-1}. To acquire more information on the electrochemical capacity of as-prepared xLFP·yLVP electrodes, the ultrahigh rate of 20 C was performed to make a critical assessment of the cyclability (Figure 3e). As expected, the LFVP electrode could deliver a higher capacity of 87.3 mA·h·g^{-1} and an energy density of 291.3 W·h·kg^{-1} after 2000 cycles. Significantly, series of xLFP·yLVP electrodes obtained good energy retentions of 77.6%, 86.3%, 88.4%, and 92.1%. Because of the high power capability of xLFP·yLVP electrodes, our effective material design strategy in this work provides an opportunity for improving the rate performance of LFP in application-oriented Li-ion batteries.

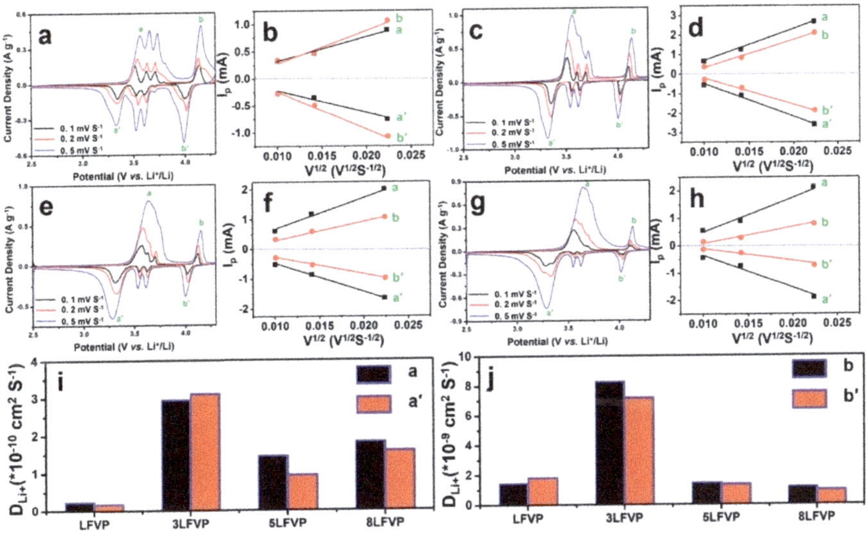

Figure 2. (**a**,**c**,**e**,**g**) CV curves of LFVP, 3LFVP, 5LFVP, and 8LFVP, respectively; (**b**,**d**,**f**,**h**) relationship between I_p and $V^{1/2}$ in LFVP, 3LFVP, 5LFVP, and 8LFVP electrodes, respectively; (**i**,**j**) Li-ion diffusion coefficients of LFP and LVP (labeled as a/a′ and b/b′) in series of xLFP·yLVP electrodes.

Compacted density is a critical parameter to evaluate whether an electrode material can be used in application-oriented Li-ion batteries. Previous reports claimed that high tap density could lead to higher gravimetric and volumetric energy densities. In other words, electrode materials with high tap density would possess high compacted density, which is not entirely precise. In fact, the compacted density can be determined by a variety of factors, involving the components of electrode pieces, the particle size distribution, and the intrinsic true density of electrode material. Hence, here, we applied a laser particle size analyzer to evaluate the size distribution of xLFP·yLVP. As shown in Figure 4a–d, the size range of 10–20 μm occupied a large proportion of the total volume for LFVP, 3LFVP, 5LFVP, and 8LFVP cathode materials, reaching up to 32.08%, 43.04%, 49.17%, and 47.17% of the total, respectively. It is worth noting that 5LFVP with the D_{50} diameter of 18.66 μm demonstrated an optimal size distribution, which may remedy the interstitial space between particles and achieve an enhanced supporting effect [41]. As shown in Figure 4e, 15 tons of pressure imposed on the electrode pieces could compress the invalid space to make close contacts between particles. For the same mass loading, the higher compacted density indicated a thinner electrode and shortened electron pathway. Moreover, Figure 4f attempts to reveal the relation between the compacted and tap density of xLFP·yLVP electrode pieces. 3LFVP provided the highest tap density of 1.08 g·cm^{-3}, but a compacted density of only 1.64 g·cm^{-3}. The equivalent two values of 5LFVP were 1.03 and 1.82 g·cm^{-3}, respectively. According to these results, the tap density of an electrode material does not have a distinct impact on the compacted density of electrode pieces, which can be deeply influenced by size distribution.

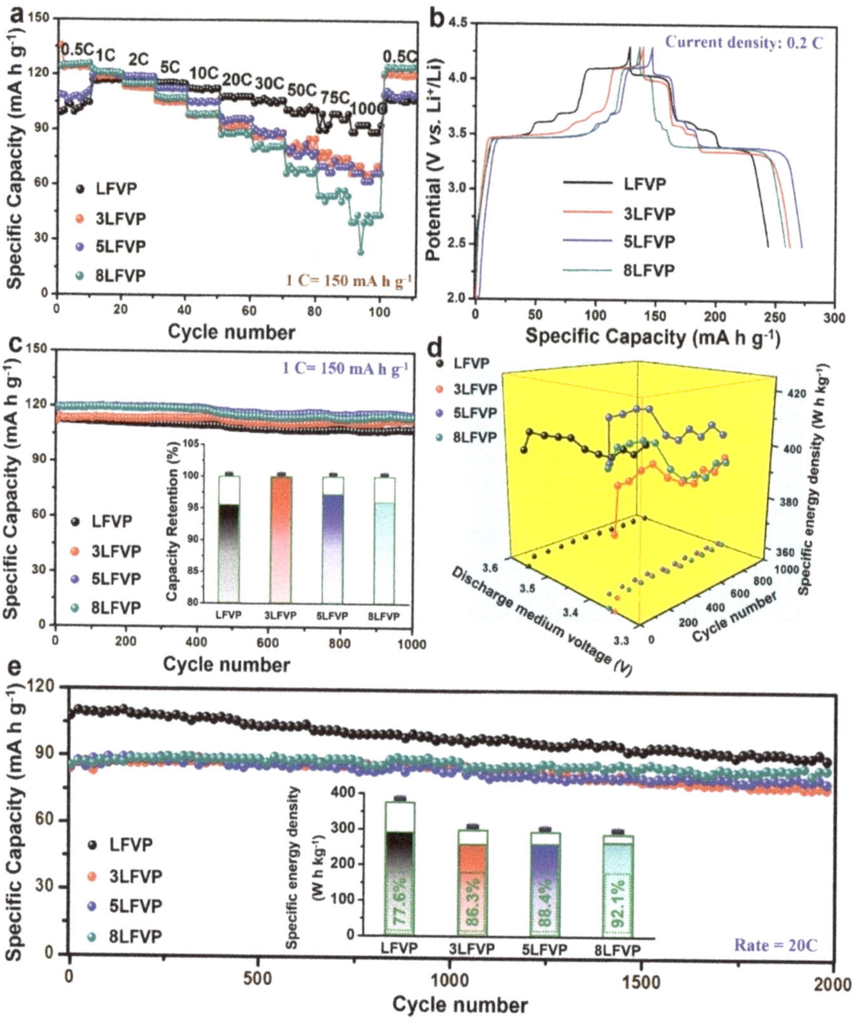

Figure 3. The comparison of xLFP·yLVP electrodes in terms of Li-ion storage capability estimated in the range 2.5–4.3 V; (**a**) rate performance; (**b**) the galvanostatic charge–discharge curves; (**c**) cycling performance at 1 C; (**d**) comparison of the material energy density; (**e**) cycling performance at the high rate of 20 C.

A concise evaluation of four types of xLFP·yLVP cathode materials was performed on the basis of six aspects: industrial feasibility, cost of raw material, energy density, power density, cycling performance, and machining capability. As shown in Figure 4g, the increased proportion of LVP in xLFP·yLVP would increase the raw cost, impeding future application in practical Li-ion batteries. Therefore, considering the manufacturing route, electrochemical performance, machining capability, and cost of series of xLFP·yLVP composites, 5LFVP is suggested to be an optimal choice for Li-ion batteries.

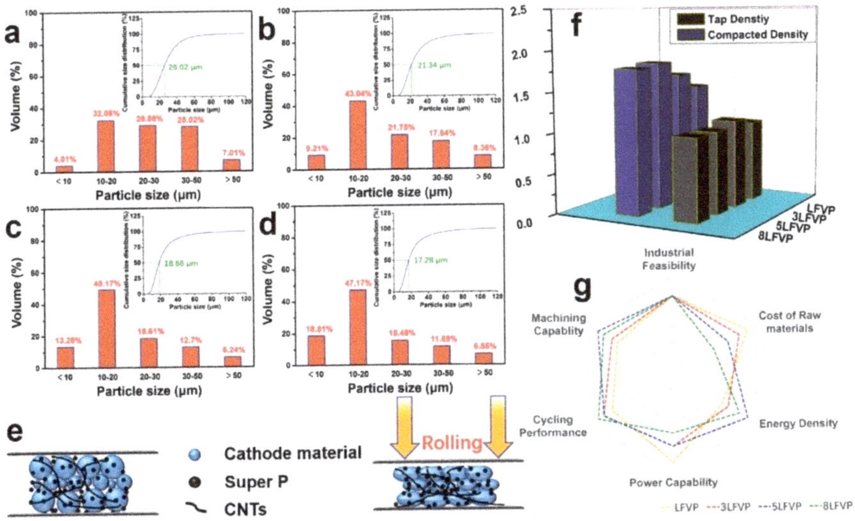

Figure 4. (**a–d**) Size distribution of xLFP·yLVP electrode materials, corresponding to LFVP, 3LFVP, 5LFVP, and 8LFVP, respectively; (**e**) illustration of the compaction process of electrode pieces; (**f**) comparison of xLFP·yLVP in terms of tap density for materials and compacted density for electrode pieces; (**g**) compressive evaluation of the series of xLFP·yLVP electrode materials.

Commercialized $Li_4Ti_5O_{12}$ with a Spinel structure exhibits superior Li-ion storage capability with long cycling stability at a high rate [42,43], ascribed to the negligible volume change during the Li-ion extraction/insertion process. Therefore, LTO was selected as an anode material in our expected prototype of Li-ion full cell. As shown in Figure S3, commercial LTO with a pure phase exhibited the morphology of block-shaped particles with an average diameter of 1 μm. For achieving a balance between cathode and anode capacity, an LTO| |Li half-cell was assembled to estimate the specific capacity and cyclability of LTO electrodes in Figure S4.

Here, the Li-ion full battery was assembled with 5LFVP as the cathode and LTO as the anode. The electrochemical behavior of 5LFVP| |LTO was analyzed by the CV test, in which the scan rate was set at 0.1 mV·s^{-1}. Four couples of redox peaks can be clearly observed from the CV curve in the potential range of 1.2–2.8 V (Figure 5a), owing to the reversible intercalation/deintercalation behavior of Li^+ ions between the 5LFVP cathode and LTO anode electrodes. It is well documented that assembling Li-ion pouch cells is a more rational way to demonstrate the potential of electrode materials in practical LIBs. As shown in Figure 5b, our 5LFVP| |LTO Li-ion pouch cell presented an excellent energy storage capability. At the constant current charge/discharge test of 400 mA (Figure 5c), 75% of the initial capacity (379.1 mA·h) was still achieved after 300 cycles. Significantly, the Coulombic efficiency remained steady at a level of around 100%, indicating great electrochemical reversibility. Moreover, under the test conditions of charging at 200 mA and discharging at 1200 mA (Figure 5d), the initial capacity could reach up to 360.3 mA h. When the cycle number increased to 300 cycles, a capacity retention of 60.3% was recorded for the pouch cell with the capacity fading by 0.15% per cycle. Although the electrochemical performance of 5LFVP| |LTO Li-ion pouch cells does not yet meet the requirements for commercialization, this prototype provides a reasonable and scientific attempt for promoting the development of xLFP·yLVP composites in application-oriented Li-ion batteries.

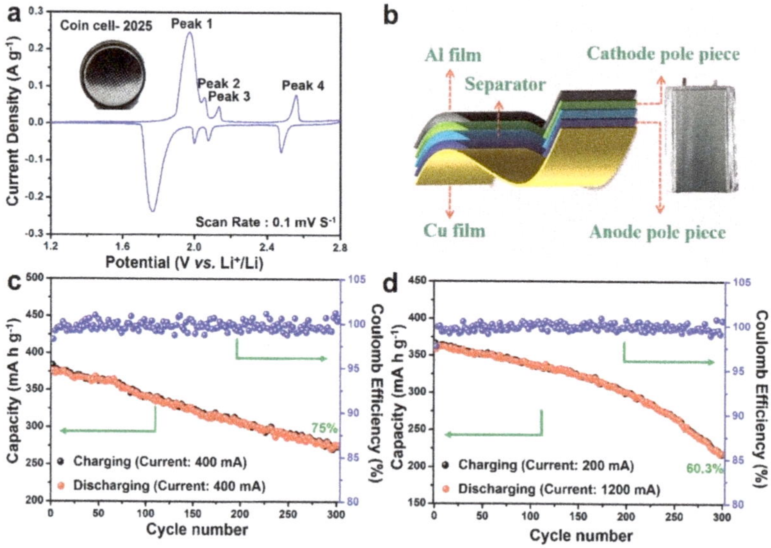

Figure 5. (**a**) CV curves of the 5LFVP||LTO Li-ion full cell measured in the potential range of 1.2–2.8 V; (**b**) schematic of 5LFVP||LTO Li-ion pouch cells; (**c**) cycling performance of 5LFVP||LTO Li-ion pouch cells at the constant current of 400 mA with the capacity retention (**d**) cycling performance of 5LFVP||LTO Li-ion pouch cells under the test conditions of charging at 200 mA and discharging at 1200 mA with the capacity retention.

6. Conclusions

In this paper, material engineering optimization was performed to prepare a series of xLFP·yLVP composites using an uncomplicated spray-drying approach, including LFVP, 3LFVP, 5LFVP, and 8LFVP. Taking all the scientific and practical factors into account, 5LFP·LVP stands out as a strong candidate cathode material in Li-ion batteries, attributed to its excellent power capability with a material energy density of 413.6 W·h·kg^{-1}, great machining capability with a compacted density of 1.82 g·cm^{-3}, and lower raw material cost consumption. Furthermore, the application-oriented energy storage device of the 5LFVP||LTO pouch cell was exploited to demonstrate the Li-ion storage capability. At both low discharge currents of 400 mA and higher currents of 1200 mA, it exhibited superior electrochemical properties. Significantly, the Coulombic efficiency could be maintained around 100% throughout the charge/discharge progress, implying a great electrochemical reversibility. This prototype provides a reasonable and scientific attempt to promote the development of xLFP·yLVP composites in application-oriented Li-ion batteries.

Supplementary Materials: The following supporting information can be downloaded at https://www.mdpi.com/article/10.3390/ma15103668/s1: Figure S1: The XRD patterns of LFP and LVP, Figure S2. (a–c) SEM images of LFVP/C, 3LFVP/C and 5LFVP/C, respectively; Figure S3. (a) XRD pattern of LTO; (b) SEM image of LTO; (c,d) TEM and HRTEM images of LTO; Figure S4. Li-ion storage capability of LTO estimated in the 0.8–2.2 V; (a) CV curve; (b) The galvanostatic charge–discharge curves at 0.2 C; (c) Rate performance; (d,e) Cycling performance at the rate of 1 and 5 C, respectively; Table S1. The theoretical content of LFP and LVP in xLFP·yLVP composites.

Author Contributions: Conceptualization, Y.P., W.X., Y.D. and F.W.; Investigation, Y.P., G.L. and P.W.; Methodology, W.X., J.Y. and X.Z.; Writing—original draft, Y.P.; Writing—review and editing, P.W., W.X., J.Y., X.Z., Z.F., X.Y., L.W. and F.W.; Formal analysis, Y.P., G.L., Z.F., X.Y. and L.W.; Supervision, Y.D. All authors have read and agreed to the published version of the manuscript.

Funding: This research was funded by the National Nature Science Foundation of China (51402096), the Educational Commission of Hubei Province of China (Q20212705), and the Nature Science Foundation of Hubei Province of China (T201517).

Institutional Review Board Statement: Not applicable.

Informed Consent Statement: Not applicable.

Data Availability Statement: Not applicable.

Conflicts of Interest: The authors declare no conflict of interest.

References

1. Broadstock, D.; Ji, Q.; Managi, S.; Zhang, D.Y. Pathways to carbon neutrality: Challenges and opportunities. *Resour. Conserv. Recycl.* **2021**, *169*, 105472. [CrossRef]
2. Chen, S.Q.; Fang, K.; Dhakal, S.; Kharrazi, A.; Tong, K.K.; Ramaswami, A. Reshaping urban infrastructure for a carbon-neutral and sustainable future. *Resour. Conserv. Recycl.* **2021**, *174*, 105765. [CrossRef]
3. Yang, Z.G.; Zhang, J.L.; Kintner-Meyer, M.C.W.; Lu, X.C.; Choi, D.W.; Lemmon, J.P.; Liu, J. Electrochemical Energy Storage for Green Grid. *Chem. Rev.* **2011**, *111*, 3577–3613. [CrossRef] [PubMed]
4. Chu, S.; Majumdar, A. Opportunities and challenges for a sustainable energy future. *Nature* **2012**, *488*, 294–303. [CrossRef]
5. Xiong, F.Y.; Jiang, Y.L.; Cheng, L.; Yu, R.H.; Tan, S.S.; Tang, C.; Zuo, C.L.; An, Q.Y.; Zhao, Y.L.; Aaumet, J.-J.; et al. Low-strain TiP_2O_7 with three-dimensional ion channels as long-life and high-rate anode material for Mg-ion batteries. *Interdiscipl. Mater.* **2022**, *1*, 140–147. [CrossRef]
6. Tarascon, J.M.; Armand, M. Issues and challenges facing rechargeable lithium batteries. *Nature* **2001**, *414*, 359–367. [CrossRef]
7. Tang, Y.X.; Zhang, Y.Y.; Li, W.L.; Ma, B.; Chen, X.D. Rational material design for ultrafast rechargeable lithium-ion batteries. *Chem. Soc. Rev.* **2015**, *44*, 5926–5940. [CrossRef]
8. Goodenough, J.B.; Park, K.S. The Li-Ion Rechargeable Battery: A Perspective. *J. Am. Chem. Soc.* **2013**, *135*, 1167–1176. [CrossRef]
9. Tan, H.T.; Xu, L.H.; Geng, H.B.; Rui, X.H.; Li, C.C.; Huang, S.M. Nanostructured $Li_3V_2(PO_4)_3$ Cathodes. *Small* **2018**, *14*, 1800567. [CrossRef]
10. Ling, J.; Karuppiah, C.; Krishnan, S.G.; Reddy, M.V.; Misnon, I.I.; Ab Rahim, M.H.; Yang, C.C.; Jose, R. Phosphate Polyanion Materials as High-Voltage Lithium-Ion Battery Cathode: A Review. *Energy Fuel* **2021**, *35*, 10428–10450. [CrossRef]
11. Zhang, Y.; Zhang, Z.H.; Tang, Y.K.; Jia, D.Z.; Huang, Y.D.; Pang, W.K.; Guo, Z.P.; Zhou, Z. $LiFePO_4$ Particles Embedded in Fast Bifunctional Conductor rGO&C@$Li_3V_2(PO_4)_3$ Nanosheets as Cathodes for High-Performance Li-Ion Hybrid Capacitors. *Adv. Funct. Mater.* **2019**, *29*, 1807895.
12. Hu, J.T.; Huang, W.Y.; Yang, L.Y.; Pan, F. Structure and performance of the $LiFePO_4$ cathode material: From the bulk to the surface. *Nanoscale* **2020**, *12*, 15036–15044. [CrossRef] [PubMed]
13. Wang, J.J.; Sun, X.L. Understanding and recent development of carbon coating on $LiFePO_4$ cathode materials for lithium-ion batteries. *Energy Environ. Sci.* **2012**, *5*, 5163–5185. [CrossRef]
14. Wang, X.F.; Feng, Z.J.; Hou, X.L.; Liu, L.L.; He, M.; He, X.S.; Huang, J.T.; Wen, Z.H. Fluorine doped carbon coating of $LiFePO_4$ as a cathode material for lithium-ion batteries. *Chem. Eng. J.* **2020**, *379*, 122371. [CrossRef]
15. Rui, X.H.; Yan, Q.Y.; Skyllas-Kazacos, M.; Lim, T.M. $Li_3V_2(PO_4)_3$ cathode materials for lithium-ion batteries: A review. *J. Power Sources* **2014**, *258*, 19–38. [CrossRef]
16. Zhang, X.F.; Kuhnel, R.S.; Hu, H.T.; Eder, D.; Balducci, A. Going nano with protic ionic liquids-the synthesis of carbon coated $Li_3V_2(PO_4)_3$ nanoparticles encapsulated in a carbon matrix for high power lithium-ion batteries. *Nano Energy* **2015**, *12*, 207–214. [CrossRef]
17. Liang, S.Q.; Cao, X.X.; Wang, Y.P.; Hu, Y.; Pan, A.Q.; Cao, G.Z. Uniform $8LiFePO_4 \cdot Li_3V_2(PO_4)_3/C$ nanoflakes for high-performance Li-ion batteries. *Nano Energy* **2016**, *22*, 48–58. [CrossRef]
18. Zhang, J.F.; Shen, C.; Zhang, B.; Zheng, J.C.; Peng, C.L.; Wang, X.W.; Yuan, X.B.; Li, H.; Chen, G.M. Synthesis and performances of $2LiFePO_4 \cdot Li_3V_2(PO_4)_3/C$ cathode materials via spray drying method with double carbon sources. *J. Power Sources* **2014**, *267*, 227–234. [CrossRef]
19. Jo, J.; Gim, J.; Song, J.; Kim, Y.; Mathew, V.; Kim, S.; Kim, S.; Park, S.; Baboo, J.P.; Kim, J. One-pot pyro-synthesis of a high energy density $LiFePO_4$-$Li_3V_2(PO_4)_3$ nanocomposite cathode for lithium-ion battery applications. *Ceram. Int.* **2017**, *43*, 4288–4294. [CrossRef]
20. He, W.; Wei, C.L.; Zhang, X.D.; Wang, Y.Y.; Liu, Q.Z.; Shen, J.X.; Wang, L.Z.; Yue, Y.Z. $Li_3V_2(PO_4)_3/LiFePO_4$ composite hollow microspheres for wide voltage lithium ion batteries. *Electrochim. Acta* **2016**, *219*, 682–692. [CrossRef]
21. Shi, Y.; Wen, L.; Li, F.; Cheng, H.M. Nanosized $Li_4Ti_5O_{12}$/graphene hybrid materials with low polarization for high rate lithium ion batteries. *J. Power Sources* **2011**, *196*, 8610–8617. [CrossRef]
22. Bruce, P.G.; Scrosati, B.; Tarascon, J.M. Nanomaterials for rechargeable lithium batteries. *Angew. Chem. Int. Ed.* **2008**, *47*, 2930–2946. [CrossRef] [PubMed]
23. Chen, L.; Chen, Z.Y.; Liu, S.S.; Gao, B.F.; Wang, J.W. Effects of Particle Size Distribution on Compacted Density of Lithium Iron Phosphate 18650 Battery. *J. Electrochem. Energy* **2018**, *15*, 41011. [CrossRef]

24. Fey, G.T.K.; Lin, Y.C.; Kao, H.M. Characterization and electrochemical properties of high tap-density $LiFePO_4/C$ cathode materia by a combination of carbothermal reduction and molten salt methods. *Electrochim. Acta* **2012**, *80*, 41–49. [CrossRef]
25. Wang, J.S.; Shen, Z.G.; Yi, M. Hydraulic Compaction on Electrode to Improve the Volumetric Energy Density of $LiFePO_4$/Graphi Batteries. *Ind. Eng. Chem. Res.* **2019**, *58*, 15407–15415. [CrossRef]
26. Yi, D.W.; Cui, X.M.; Li, N.L.; Zhang, L.; Yang, D.Y. Enhancement of Electrochemical Performance of $LiFePO_4$@C by Ga Coatir *ACS Omega* **2020**, *5*, 9752–9758. [CrossRef]
27. Zhang, Q.; Huang, S.Z.; Jin, J.; Liu, J.; Li, Y.; Wang, H.E.; Chen, L.H.; Wang, B.J.; Su, B.L. Engineering 3D bicontinuous hierarchical macro-mesoporous $LiFePO_4/C$ nanocomposite for lithium storage with high rate capability and long cycle stability. *Sci. Re* **2016**, *6*, 25942. [CrossRef]
28. Luo, Y.Z.; Xu, X.; Zhang, Y.X.; Pi, Y.Q.; Zhao, Y.L.; Tian, X.C.; An, Q.Y.; Wei, Q.L.; Mai, L.Q. Hierarchical Carbon Decorate $Li_3V_2(PO_4)_3$ as a Bicontinuous Cathode with High-Rate Capability and Broad Temperature Adaptability. *Adv. Energy Mater.* **201** *4*, 1400107. [CrossRef]
29. Cheng, Y.; Feng, K.; Wang, H.Q.; Zhang, H.Z.; Li, X.F.; Zhang, H.M. "Three-in-One": A New 3D Hybrid Structure of $Li_3V_2(PO_4)_3$ Biomorphic Carbon for High-Rate and Low-Temperature Lithium Ion Batteries. *Adv. Mater. Interfaces* **2017**, *4*, 1700686. [CrossRe
30. Wu, X.L.; Jiang, L.Y.; Cao, F.F.; Guo, Y.G.; Wan, L.J. $LiFePO_4$ Nanoparticles Embedded in a Nanoporous Carbon Matrix: Superic Cathode Material for Electrochemical Energy-Storage Devices. *Adv. Mater.* **2009**, *21*, 2710–2714. [CrossRef]
31. Li, Z.; He, L.H.; Zhu, Y.F.; Yang, C. A Green and Cost-Effective Method for Production of LiOH from Spent $LiFePO_4$. *ACS Sustai Chem. Eng.* **2020**, *8*, 15915–15926. [CrossRef]
32. Cui, X.M.; Yi, D.W.; Li, N.L.; Zhang, L.; Zhang, X.F.; Yang, D.Y. Novel $LaFeO_3$ Coating Modification for a $LiFePO_4$ Cathod *Energy Fuel* **2020**, *34*, 7600–7606. [CrossRef]
33. Peng, Y.; Tan, R.; Ma, J.M.; Li, Q.H.; Wang, T.H.; Duan, X.C. Electrospun $Li_3V_2(PO_4)_3$ nanocubes/carbon nanofibers as fre standing cathodes for high-performance lithium-ion batteries. *J. Mater. Chem. A* **2019**, *7*, 14681–14688. [CrossRef]
34. Ding, X.K.; Zhang, L.L.; Yang, X.L.; Fang, H.; Zhou, Y.X.; Wang, J.Q.; Ma, D. Anthracite-Derived Dual-Phase Carbon-Coate $Li_3V_2(PO_4)_3$ as High-Performance Cathode Material for Lithium Ion Batteries. *ACS Appl. Mater. Interfaces* **2017**, *9*, 42788–4279 [CrossRef]
35. Sun, H.X.; Du, H.R.; Yu, M.K.; Huang, K.F.; Yu, N.; Geng, B.Y. Vesicular $Li_3V_2(PO_4)_3/C$ hollow mesoporous microspheres as a efficient cathode material for lithium-ion batteries. *Nano Res.* **2019**, *12*, 1937–1942. [CrossRef]
36. Wang, B.; Zhang, Z.K.; Ning, Y.; Li, X.K.; Ruan, T.T.; Wang, F.; Wang, D.L.; Zhou, Y. Construction of Dual-Carbon Co-Modifie $LiFePO_4$ Nanocrystals via Microreactor Strategy for High-Performance Lithium Ion Batteries. *Energy Technol.* **2020**, *8*, 200017 [CrossRef]
37. Zhao, Y.; Peng, L.L.; Liu, B.R.; Yu, G.H. Single-Crystalline $LiFePO_4$ Nanosheets for High-Rate Li-Ion Batteries. *Nano Lett.* **2014**, *1* 2849–2853. [CrossRef]
38. Ding, M.L.; Cheng, C.; Wei, Q.L.; Hu, Y.; Yan, Y.Y.; Dai, K.H.; Mao, J.; Guo, J.H.; Zhang, L.; Mai, L.Q. Carbon decorate $Li_3V_2(PO_4)_3$ for high-rate lithium-ion batteries: Electrochemical performance and charge compensation mechanism. *J. Energ Chem.* **2021**, *53*, 124–131. [CrossRef]
39. Pan, A.Q.; Choi, D.W.; Zhang, J.G.; Liang, S.Q.; Cao, G.Z.; Nie, Z.M.; Arey, B.W.; Liu, J. High-rate cathodes based on $Li_3V_2(PO_4)$ nanobelts prepared via surfactant-assisted fabrication. *J. Power Sources* **2011**, *196*, 3646–3649. [CrossRef]
40. Patil, V.; Oh, W.; Yoo, J.W.; Pu, L.; Park, J.H.; Yoon, W.S.; Yi, G.R. Carbon-Coated Supraballs of Randomly Packed LiFePO Nanoplates for High Rate and Stable Cycling of Li-Ion Batteries. *Part. Part. Syst. Charact.* **2019**, *36*, 1900149. [CrossRef]
41. Xu, Q.; Li, J.Y.; Sun, J.K.; Yin, Y.X.; Wan, L.J.; Guo, Y.G. Watermelon-Inspired Si/C Microspheres with Hierarchical Buffe Structures for Densely Compacted Lithium-Ion Battery Anodes. *Adv. Energy Mater.* **2017**, *7*, 1601481. [CrossRef]
42. Xu, H.H.; Hu, X.L.; Sun, Y.M.; Luo, W.; Chen, C.J.; Liu, Y.; Huang, Y.H. Highly porous $Li_4Ti_5O_{12}/C$ nanofibers for ultrafas electrochemical energy storage. *Nano Energy* **2014**, *10*, 163–171. [CrossRef]
43. Yu, L.; Wu, H.B.; Lou, X.W. Mesoporous $Li_4Ti_5O_{12}$ Hollow Spheres with Enhanced Lithium Storage Capability. *Adv. Mater.* **2013** *25*, 2296–2300. [CrossRef] [PubMed]

materials MDPI

Article

Oxygen-Plasma-Induced Hetero-Interface $NiFe_2O_4/NiMoO_4$ Catalyst for Enhanced Electrochemical Oxygen Evolution

Nuo Xu [1,†], Wei Peng [2,†], Lei Lv [2], Peng Xu [2], Chenxu Wang [1], Jiantao Li [3], Wen Luo [1,2,*] and Liang Zhou [2]

1 Department of Physics, School of Science, Wuhan University of Technology, Wuhan 430070, China; mr.promise@whut.edu.cn (N.X.); 245856@whut.edu.cn (C.W.)
2 State Key Laboratory of Advanced Technology for Materials Synthesis and Processing, Wuhan University of Technology, Wuhan 430070, China; pw0225@whut.edu.cn (W.P.); lvlei1028@whut.edu.cn (L.L.); xupeng620@whut.edu.cn (P.X.); liangzhou@whut.edu.cn (L.Z.)
3 Chemical Sciences and Engineering Division, Argonne National Laboratory, Lemont, IL 60439, USA; jiantao_work@126.com
* Correspondence: luowen_1991@whut.edu.cn
† These authors contributed equally to this work.

Abstract: The electrolysis of water to produce hydrogen is an effective method for solving the rapid consumption of fossil fuel resources and the problem of global warming. The key to its success is to design an oxygen evolution reaction (OER) electrocatalyst with efficient conversion and reliable stability. Interface engineering is one of the most effective approaches for adjusting local electronic configurations. Adding other metal elements is also an effective way to enrich active sites and improve catalytic activity. Herein, high-valence iron in a heterogeneous interface of $NiFe_2O_4/NiMoO_4$ composite was obtained through oxygen plasma to achieve excellent electrocatalytic activity and stability. In particular, 270 mV of overpotential is required to reach a current density of 50 mA cm^{-2}, and the overpotential required to reach 500 mA cm^{-2} is only 309 mV. The electron transfer effect for high-valence iron was determined by X-ray photoelectron spectroscopy (XPS). The fast and irreversible reconstruction and the true active species in the catalytic process were identified by in situ Raman, ex situ XPS, and ex situ transmission electron microscopy (TEM) measurements. This work provides a feasible design guideline to modify electronic structures, promote a metal to an active oxidation state, and thus develop an electrocatalyst with enhanced OER performance.

Keywords: hetero-interface; oxygen plasma; oxygen evolution reaction; electrocatalysis; electronic modulation

1. Introduction

Nowadays, the dramatically increased demand for fossil energy has resulted in the depletion of traditional energy materials and has generated concerns regarding energy security and the environmental pollution caused by the extensive use of fossil energy [1]. Overall, water electrolysis has developed as an ideal and effective approach for producing hydrogen, an alternative clean energy source to traditional fossil fuels [2,3]. The oxygen evolution reaction (OER) involves multiple steps of proton coupling and a complex four-electron transfer process [4–6]. The sluggish reaction kinetics eventually cause a high enough overpotential to trigger the OER, which is a key factor limiting the efficiency of water electrolysis [3,7]. To date, the most effective catalysts for the OER have been found to be rare metal oxides, such as IrO_2 and RuO_2, as they significantly reduce the overpotential required for the OER. However, the high expenditure and scarcity of precious metals restrict their wide application for practical industrialization as efficient electrocatalysts.

Due to the abundant transition metal resources of the Earth itself, the construction of non-noble metal OER electrocatalysts has received much attention recently [6,8–10]. This includes transition metal compounds based on nickel, iron, or cobalt (Ni, Fe, and Co nonoxides [11–14]; oxides [15–18]; hydroxides [19–21]; and oxyhydroxides [22–25]), which have shown high conversion efficiencies towards the OER as substitutes for a precious metal electrocatalyst. Meanwhile, growing nanostructured catalysts directly on conductive substrates, such as Ni foam, has been established to decrease contact resistance and effectively improve energy efficiency [26–28]. Among these transition metal compounds, $NiMoO_4$ on Ni foam has been researched as a promising electrocatalyst for its facile synthesis in large quantities. However, its intrinsic activity still remains defective, especially for its deficient active site (only Ni site) [29–32].

Generally, it is crucial to consider the modification of structure and electronic configuration in order to achieve outstanding OER performance, especially in order to attain a higher current at a lower overpotential with long-term stability [5]. An atomic-scale approach for constructing a reliable interface, especially a hetero-interface between different nanomaterials, has been intensively adopted to modify the local electronic structure of materials [33]. This approach can accelerate the reaction kinetics by combining the structural advantages of each component, thus improving the catalytic performance of nanocomposites [34–38]. For example, Lv et al. synthesized a core–shell structure of NiFe-60/Co_3O_4 on Ni Foam with an obvious and clear hetero-interface between the Co_3O_4 nanowire and the NiFe-layered double hydroxide nanosheet [39]. A hetero-interface contributes to the interaction between two different nanomaterials, facilitates electron transfer, and further leads to enhanced catalytic activity for the OER. Zhang et al. demonstrated a CoN_4-based metal−organic framework (MOF) with embedded $CoFeO_x$ nanoparticles; Co sites anchoring on the $CoFeO_x$/MOF interface brought about an altered 3D electronic configuration for the interfacial Co and a higher valence [40]. In addition, composites consisting of multivariate transition metals can promote the exceptional modification of active sites within the matrix, and thus improve reaction efficiency and durability [41]. Based on this, the addition of Fe elements has been confirmed as a proper approach to enrich the active sites and boost highly efficient OER performance [42–44].

Herein, we report a hetero-interface made of $NiFe_2O_4$ nanoparticles (NPs) and a $NiMoO_4$ nanowire (denoted as $NiFe_2O_4/NiMoO_4$). Briefly, $NiMoO_4$ nanowires were prepared on nickel foam through a facile hydrothermal synthesis. NiFe Prussian blue analogs (NiFe PBA) were firstly fixed on the $NiMoO_4$ nanowires by iron exchange. Then, oxygen (O_2) plasma converted the NiFe PBA to $NiFe_2O_4$ to form a $NiFe_2O_4/NiMoO_4$ hetero-interface. As-synthesized, the $NiFe_2O_4/NiMoO_4$ exhibits excellent performance for the OER with a low overpotential of 309 mV required to reach 500 mA cm^{-2} and a satisfactory stability (a 4% increase in the overpotential at 50 mA cm^{-2} over 150 h). The shift in the binding energy of metal sites increased the electronic interaction of the modulated hetero-interface. To understand its excellent OER performance, in situ Raman measurements, ex situ transmission electron microscopy (TEM), and ex situ X-ray photoelectron spectroscopy (XPS) were used to confirm the fast and irreversible reconstruction and identify the true active species in the catalytic process.

2. Results and Discussion

The schematics shown in Figure 1a illustrate the preparation of $NiFe_2O_4$ nanoparticles integrated with $NiMoO_4$ nanowires on nickel foam. Briefly, through a simple hydrothermal method [45], hydrated $NiMoO_4$ nanowires were vertically germinated on Ni foam. In accordance with a previous report [46], the NiFe PBA was grown on $NiMoO_4$ nanowires and the MoO_4^{2-} on the surface of the nanowires was replaced with K^+ and $[Fe(CN)_6]^{3+}$ by ion exchange. The NiFe PBA on the $NiMoO_4$ surface was converted to $NiFe_2O_4$ NPs under O_2 plasma treatment, and $NiFe_2O_4/NiMoO_4$ was obtained. For comparison, $NiMoO_4$ nanowires were also placed under the same O_2 plasma treatment and denoted as $NiMoO_4$ O_2-Pl.

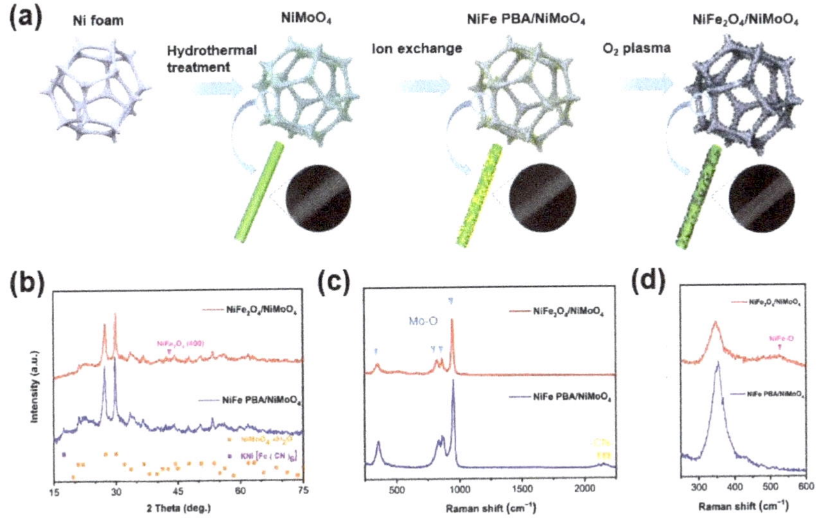

Figure 1. (**a**) Schematic illustration of the synthesis of $NiFe_2O_4/NiMoO_4$; (**b**) XRD patterns of NiFe PBA/$NiMoO_4$ and $NiFe_2O_4/NiMoO_4$; (**c**) Raman spectra of NiFe PBA/$NiMoO_4$ and $NiFe_2O_4/NiMoO_4$; and (**d**) Raman spectra of NiFe PBA/$NiMoO_4$ and $NiFe_2O_4/NiMoO_4$ in a region from 250 cm^{-1} to 750 cm^{-1}, respectively.

Characterization of $NiMoO_4$. Figure S1a,b indicate that the $NiMoO_4$ possessed an even and well-defined nanowire morphology. Its average diameter was about 100 nm. As the XRD patterns show in Figure S2a, the diffraction peaks of the $NiMoO_4$ were in perfect agreement with the $NiMoO_4 \cdot xH_2O$ (JCPDF: 00-013-0128), which means the $NiMoO_4 \cdot xH_2O$ was highly crystalline. In the FT-IR spectra shown in Figure S3, the two peaks at 1628 and 3446 cm^{-1} correspond to the stretching vibration of hydroxyl (-OH), which can be ascribed to the bending mode of crystal water in the $NiMoO_4 \cdot xH_2O$ and the surface-adsorbed water molecules [47].

Characterization of NiFe PBA/$NiMoO_4$. As shown in Figure 1b, a weak diffraction peak appears at 17.3°, which can be attributed to the KNi[Fe(CN)$_6$] (JCPDF: 01-089-8978). At the same time, the diffraction peaks of the $NiMoO_4 \cdot xH_2O$ still remain in the NiFe PBA/$NiMoO_4$. In Figure S1c,d, the surface of the $NiMoO_4$ nanowires is covered with small-sized NiFe PBA NPs, indicating the expected process of the iron exchange. Figure 1c reveals peaks at 357, 828, 872, and 950 cm^{-1} for the NiFe PBA/$NiMoO_4$, which can be assigned to the Mo-O vibration, and this result is consistent with previous reports [32]. In addition, the NiFe PBA/$NiMoO_4$ exhibits three peaks around 2100 cm^{-1}, which can be attributed to -CN [48,49]. In Figure S3, for the NiFe PBA/$NiMoO_4$, a new characteristic peak can be observed at 2099 cm^{-1} in the FT-IR spectrum, which is attributed to the stretching vibrations of -CN in the NiFe PBA NPs [50]. In Figure S5a, for the NiFe PBA/$NiMoO_4$, the Ni 2p spectra can be deconvoluted into four peaks and two wide satellite peaks. In the Ni $2p_{3/2}$, the peaks at 856.2 eV and 857 eV can be ascribed to the Ni^{2+} and Ni^{3+} species, respectively. Meanwhile, in the Ni $2p_{1/2}$, the peaks of the Ni^{2+} and Ni^{3+} species can be fitted at 874.0 eV and 875.2 eV, respectively. In addition, two satellite peaks for Ni can be observed at 862.8 and 880.7 eV [51]. As shown in Figure S5b, the further fitted peaks at 708.5 and 709.1 eV in the Fe $2p_{3/2}$ are ascribed to Fe^{2+} and Fe^{3+}, respectively, while the peaks observed at 721.6 and 722.2 eV in the Fe $2p_{1/2}$ are owed to the existence of Fe^{2+} and Fe^{3+}, respectively. The ratio of the Fe^{3+} to Fe^{2+} peak areas in the Fe $2p_{3/2}$ was calculated as 0.948. Furthermore, one satellite peak for Fe appears at 715.7 eV [51].

Characterization of $NiFe_2O_4/NiMoO_4$. In Figure 1b, compared to the NiFe PBA/$NiMoO_4$, the diffraction peak initially attributed to KNi[Fe(CN)$_6$] disappears, and a new weak diffraction peak can be observed at 43.3°, which is attributed to the (400) planes of the $NiFe_2O_4$ (JCPDS: 44-1485). As shown in Figure 1c, the characteristic Raman peaks attributed to the Mo-O bond still remain, while the peak owed to -CN disappears. A broad peak at 520 cm^{-1} in Figure S2b indicates the formation of NiFe-O. The same phenomenon is shown in Figure S3, as the characteristic peak of -C≡N- disappears and an apparent peak at 1384 cm^{-1} can be assigned to the C=O in CO_2 and the stretching vibration of the interlayer $NO_3^=$ groups [47]. This is related to the decomposition of PBA under O_2 plasma treatment. As shown in Figure 2a, the $NiMoO_4$ remains in the structure of the nanowire with a diameter of 100 nm, similarly to the $NiMoO_4$ and NiFe PBA/$NiMoO_4$, while the $NiFe_2O_4$ nanoparticles slightly agglomerate. The TEM image of the $NiFe_2O_4/NiMoO_4$ (Figure 2b) clearly shows the $NiFe_2O_4$ NPs anchoring on the surface of the $NiMoO_4$ nanowire. The high-resolution transmission electron microscope (HRTEM) image in Figure 2c indicates the high-crystalline characteristic of the $NiMoO_4$. The HRTEM image in Figure 2d shows the apparent hetero-interface of the $NiFe_2O_4/NiMoO_4$. The lattice spacing of 2.08 Å can be attributed to the (400) plane of the $NiFe_2O_4$ and the lattice spacing of 3.26 Å assigned to the $NiMoO_4{\cdot}xH_2O$. The elemental mapping images (Figure 2e) indicate that the Fe element is evenly distributed on the $NiMoO_4$ nanowires in the form of nanoparticles. Table S1 shows the chemical composition of the $NiFe_2O_4/NiMoO_4$. The molecular ratio of $NiFe_2O_4$ and $NiMoO_4$ in the $NiFe_2O_4/NiMoO_4$ is 1:17.27.

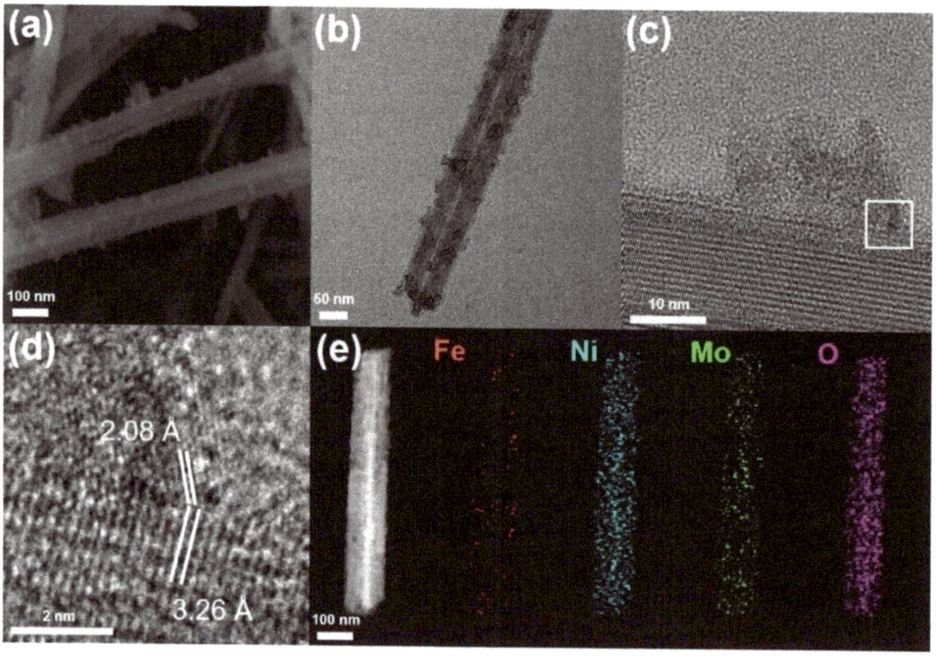

Figure 2. (**a**) SEM image, (**b**) TEM image, and (**c**) HRTEM images of $NiFe_2O_4/NiMoO_4$; (**d**) the corresponding HRTEM images of selected areas; and (**e**) EDS mapping images for Fe, Ni, Mo, and O elements of $NiFe_2O_4/NiMoO_4$.

XPS was used to analyze and further explore the surface electronic interaction of the $NiFe_2O_4/NiMoO_4$. The Ni 2p spectra of $NiFe_2O_4/NiMoO_4$ contain four fitted peaks with wide satellites (Figure 3a). The fitted peaks at 856.1 and 873.7 eV in the Ni $2p_{3/2}$ and the Ni $2p_{1/2}$, respectively, can be attributed to Ni^{2+}, while another two peaks (858.3 and 875.9 eV) correspond to the Ni^{3+} species. The wide peaks observed at 862.3 and 879.9 eV are owed to the satellites of Ni [52]. As Figure 3b shows, the fitted peaks at 710.7 and 713.3 eV can be related to Fe^{2+} and Fe^{3+} in the Fe $2p_{3/2}$, and the same is true for another two peaks at 723.8 and 726.4 eV in the Fe $2p_{1/2}$ [53]. The broad peaks at 718.8 and 731.9 eV can be attributed to the satellite peaks of Fe. In Figure S6a,b, the peaks in the Ni $2p_{3/2}$ and the Mo 3D of the $NiFe_2O_4/NiMoO_4$ exhibit slightly negative shifts of about 0.2 eV compared with those observed from the spectra of the NiFe PBA/$NiMoO_4$. The peaks of the Fe 2p exhibit a distinct positive shift compared with those of the NiFe PBA/$NiMoO_4$, and the ratio of the Fe^{3+} to Fe^{2+} peak area in the Fe $2p_{3/2}$ (Figures S5b and 3b) increases from 0.948 in the NiFe PBA/$NiMoO_4$ to 1.706 in the $NiFe_2O_4/NiMoO_4$. The negative movement of binding Ni and Mo energy indicates the regulation of the electronic structure in the hetero-interface. Meanwhile, oxygen plasma leads to the oxidation of Fe, and these two factors eventually promote an increase in the binding energy of Fe. It has been confirmed that Fe with a high valence state promotes processes in the OER [54–56]. The fitted O 1s peaks at 530.7, 532.3, and 533.3 eV can be attributed to metal-O, C=O [57], and surface-adsorbed oxygen, respectively.

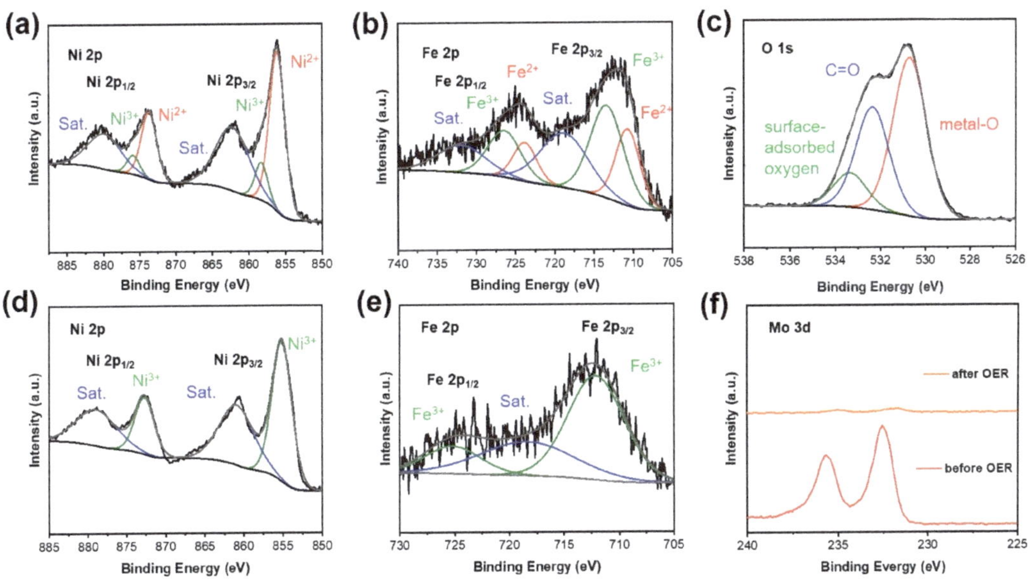

Figure 3. XPS: (**a**) Ni 2p, (**b**) Fe 2p, and (**c**) O 1s spectra of $NiFe_2O_4/NiMoO_4$; (**d**) Ni 2p and (**e**) Fe 2p spectra of $NiFe_2O_4/NiMoO_4$ after OER testing for 3 h; (**f**) Mo 3D spectra of $NiFe_2O_4/NiMoO_4$ before OER testing and after OER testing.

Electrochemical performance. For the purpose of measuring the electrochemical performance of the prepared samples, a three-electrode electrochemical system was used. An aqueous solution of 1 M KOH was selected as the electrolyte solution. The polarization curves of all samples with *iR* corrected are shown as Figure 4a. The peaks around 1.38 V can be assigned to the oxidation of nickel species to a higher valence state. Furthermore, the $NiFe_2O_4/NiMoO_4$ demonstrates the lowest overpotential of 253 mV to reach 10 mA cm^{-2}, while the overpotential required to achieve 10 mA cm^{-2} for the NiFe PBA/$NiMoO_4$, $NiMoO_4$ O_2-Pl, $NiMoO_4$, and Ni foam is 310, 313, 324, and 431 mV, respectively. In addition, for the $NiFe_2O_4/NiMoO_4$, an overpotential of 270 and 309 mV are required to achieve 50 mA cm^{-2} and 500 mA cm^{-2}, respectively.

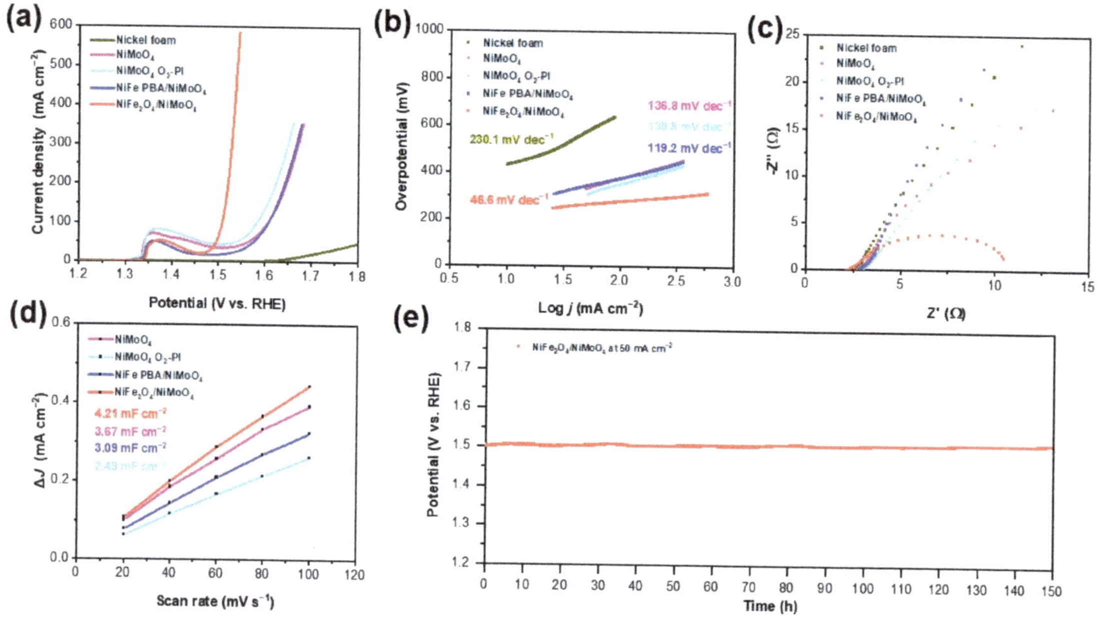

Figure 4. (**a**) Polarization curves and (**b**) corresponding Tafel slope plots of as-prepared catalysts; (**c**) EIS Nyquist plots of $NiMoO_4$, $NiMoO_4$ O_2-Pl, NiFe PBA/$NiMoO_4$, and $NiFe_2O_4/NiMoO_4$; (**d**) capacitive current densities plotted as a function of the scan rate; and (**e**) chronopotentiometry of $NiFe_2O_4/NiMoO_4$ at 50 mA cm^{-2} with *iR* corrected.

As shown in Figure 4b, the Tafel slope of the $NiFe_2O_4/NiMoO_4$ is the smallest, at 46.4 mV dec^{-1}, compared with that of the NiFe PBA/$NiMoO_4$ (119.2 mV dec^{-1}), $NiMoO_4$ O_2-Pl (139.8 mV dec^{-1}), $NiMoO_4$ (136.8 mV dec^{-1}), and Ni foam (230.1 mV dec^{-1}). The smaller Tafel slope of the $NiFe_2O_4/NiMoO_4$ indicates its faster kinetics [4,6]. The high performance of the $NiFe_2O_4/NiMoO_4$ can be attributed to the oxygen-plasma-induced formation of the hetero-interface, made up of $NiFe_2O_4$ NPs and $NiMoO_4$ nanowire arrays and containing iron with a higher valence. Iron with a higher valence has been confirmed to be conducive to the OER [54–57].

Figure 4c shows the electrochemical impedance spectroscopy (EIS) of all samples, and it can clearly be observed that the smallest charge transfer resistance (R_{ct}) is found in the $NiFe_2O_4/NiMoO_4$. The smaller R_{ct} relative to the others indicates a faster charge transfer for the $NiFe_2O_4/NiMoO_4$, which may relate to the hetero-interface made up of $NiFe_2O_4$ NPs and $NiMoO_4$ and further leads to enhanced electrocatalytic performance.

The electrochemical active surface area (ECSA) by CV measurement is shown in Figure S4. As shown in Figure 4d, the double-layer capacitance (C_{dl}) of the $NiFe_2O_4/NiMoO_4$, NiFe PBA/$NiMoO_4$, $NiMoO_4$ O_2-Pl, and $NiMoO_4$ was calculated to be 4.21, 3.09, 2.49, and 3.67 mF cm^{-2}. The larger value of C_{dl} indicates a higher electrocatalytic OER activity of the $NiFe_2O_4/NiMoO_4$, which is attributed to more exposed active sites related to the iron with a higher valence. Long-term stability is an important index for evaluating a catalyst. As shown in Figure 4e, the $NiFe_2O_4/NiMoO_4$ displays outstanding durability (a 4% increase in the overpotential at 50 mA cm^{-2} over a 150 h reaction).

In situ Raman spectra. To figure out the phase change and reconstruction in the OER process, the $NiFe_2O_4/NiMoO_4$ was first activated in an alkaline solution. In Figure 5a, with the increase in the applied potential from 1.18 V to 1.43 V, the intensity of characteristic peaks for Mo-O vibration decreased, which represents the dissolution of MoO_4^{2-}. Meanwhile, a small but sharp characteristic peak at 525 cm^{-1} corresponding with the Fe-O bond in FeOOH emerged with an applied potential of 1.23 V [58], which indicates the formation of FeOOH in the OER process. When the potential is applied at 1.28 V, broad peaks can be observed around 460 cm^{-1} and 520 cm^{-1}, which are related to the appearance of α-$Ni(OH)_2$ [59]. The peak becomes sharper when the applied potential arrives at 1.33 V. A broad peak occurs at 475 cm^{-1}, which can be attributed to the emergence of γ-NiOOH from α-$Ni(OH)_2$ [32,58], and the peak tends to become sharper with an applied potential at 1.43 V, while another characteristic peak of γ-NiOOH appears at 558 cm^{-1} [32,59].

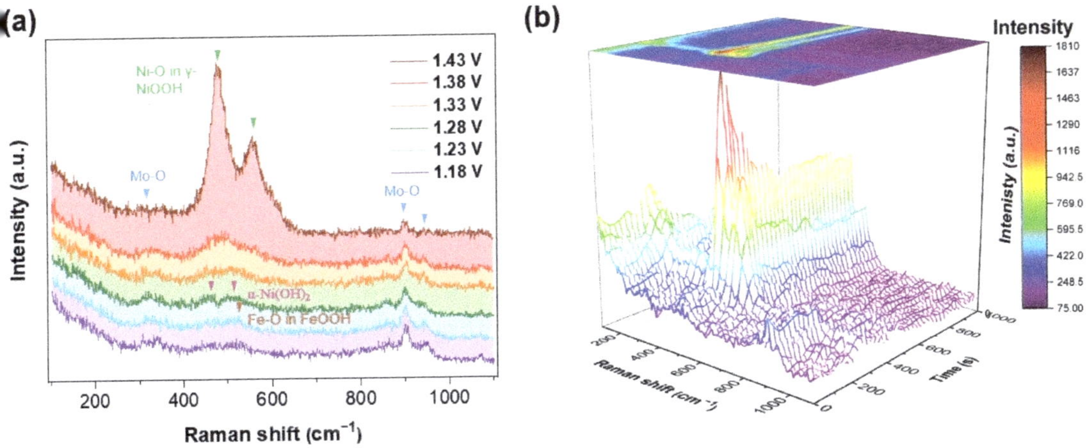

Figure 5. In situ Raman spectra of $NiFe_2O_4/NiMoO_4$ (**a**) for activation from 1.18 V to 1.43 V and (**b**) for CVs in the initial 2 cycles from 1.18 V to 1.43 V at a scan rate 1 mV s^{-1}.

In situ Raman spectra of the $NiFe_2O_4/NiMoO_4$ in the initial two cycles in CVs are shown in Figure 5b. With multiple cycles, the intensity of the characteristic peak for γ-NiOOH gradually stabilizes and the characteristic peaks of Mo-O vibration almost completely disappear, which can be attributed to the irreversible reconstruction of the $NiFe_2O_4/NiMoO_4$.

For comparison, the NiFe PBA/$NiMoO_4$ was also first activated in an alkaline solution. In Figure S8a, with the increase in applied potential, the same phenomenon of a decrease in the intensity of characteristic peaks for Mo-O vibration can be observed. In addition, when a potential of 1.38 V is applied to the $NiFe_2O_4/NiMoO_4$, a broad peak occurs at 475 cm^{-1}, which can be attributed to the emergence of γ-NiOOH. The same phenomenon occurs at an applied potential of 1.28 V for $NiFe_2O_4/NiMoO_4$. This fact means that the $NiFe_2O_4/NiMoO_4$ is reconstructed faster than the NiFe PBA/$NiMoO_4$, which leads to the better OER performance of the $NiFe_2O_4/NiMoO_4$ from another aspect. However, there are no observable peaks attributed to FeOOH, and in Figure S8b, with the increase in applied potential, the peaks related to -CN still exist [48,49]. This illustrates that the Fe coordinating with the cyanide group cannot catalyze the OER as an independent active site with the increase in applied potential, which further explains the reason that the OER performance of the NiFe PBA/$NiMoO_4$ is close to the OER performance of the $NiMoO_4$.

Ex situ XPS. The Ni 2p and Fe 2p spectra of the $NiFe_2O_4/NiMoO_4$ after 3 h of OER testing are shown in Figure 3d,e, respectively. In Figure 3d, the Ni 2p can be deconvoluted into two peaks with two satellites. The fitted peaks at 855.1 and 872.7 eV can be ascribed to Ni^{3+}, which is attributed to NiOOH [60]. Meanwhile, two satellites of Ni can be observed at 860.9 and 878.9 eV, respectively. As shown in Figure 3e, the two fitted peaks occur at 712.1 and 725.2 eV with a broad satellite at 718.3 eV, which can be related to FeOOH [61,62]. It can clearly be observed that there is a sharp attenuation in the peak intensity of the Mo 3d of the $NiFe_2O_4/NiMoO_4$ after 3 h of OER testing, further demonstrating the irreversible reconstruction of the $NiFe_2O_4/NiMoO_4$ with the dissolution of MoO_4^{2-}.

Ex situ TEM. The images of the $NiFe_2O_4/NiMoO_4$ after OER testing (Figure 6a,b) clearly show the robust surface of the nanowire and numerous defects as the result of the dissolution of MoO_4^{2-}. The HRTEM image in Figure 6c reveals clear lattice fringes of the (105) plane for NiOOH (JCPDF: 00-006-0075) with a crystalline interplanar spacing of 2.09 Å. The HRTEM image in Figure 3d shows small black particles distributed in clumps, which may relate to the amorphous FeOOH delivered by the activation of the $NiFe_2O_4$ in OER testing. The elemental mapping images (Figure 3e) indicate that Fe is still evenly distributed on the $NiMoO_4$ nanowire, and Mo dissolves in large quantities, which is consistent with the aforementioned analytical results.

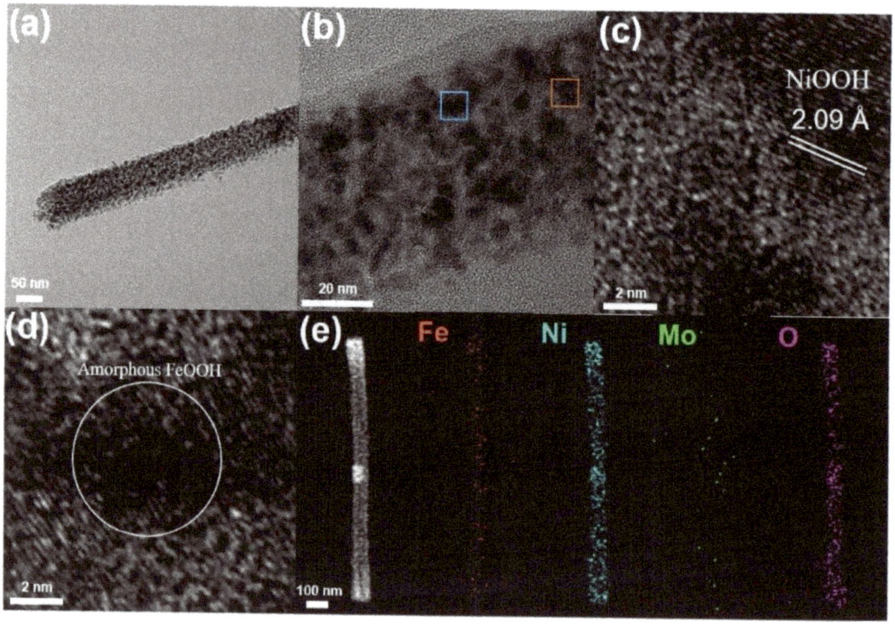

Figure 6. (**a**) TEM image; (**b**) HRTEM image of $NiFe_2O_4/NiMoO_4$ after 3 h of OER testing; (**c**) the corresponding HRTEM images of the area selected by the orange frame; (**d**) the corresponding HRTEM images of the area selected by the blue frame; and (**e**) EDS mapping images for Fe, Ni, Mo, and O elements of $NiFe_2O_4/NiMoO_4$ after OER.

3. Conclusions

In summary, a heterogeneous interface of $NiFe_2O_4/NiMoO_4$ with high-valence iron through oxygen plasma can be fabricated to achieve excellent electrocatalytic activity and stability. To achieve a current density of 50 mA cm^{-2}, 270 mV of overpotential is required, while an overpotential of 309 mV is required to reach 500 mA cm^{-2}. The $NiFe_2O_4/NiMoO_4$ also exhibits a satisfactory stability (a 4% increase in the overpotential at 50 mA cm^{-2} over 150 h). O_2-plasma-induced electronic interaction in the hetero-interface of $NiFe_2O_4/NiMoO_4$ and iron with a higher valence play an essential role in OER performance. The potential-dependent phase change and the fast and irreversible reconstruction of the $NiFe_2O_4/NiMoO_4$ in a catalytic process were identified by in situ Raman, ex situ XPS, and ex situ TEM measurements. Based on this, the true active species, NiOOH and FeOOH, were determined. This work provides a feasible design guideline for modifying electronic structure through the construction of a heterogeneous interface and the activation of metal sites by O_2 plasma, finally leading to enhanced OER performance.

Supplementary Materials: The following supporting information can be downloaded at: https://www.mdpi.com/article/10.3390/ma15103688/s1. Figure S1. SEM images (a, b) and (c, d) of $NiMoO_4$ and NiFe PBA/$NiMoO_4$, respectively. Figure S2. XRD patterns of NiFe PBA/$NiMoO_4$ and $NiFe_2O_4$/$NiMoO_4$. Figure S3. FT-IR spectra of $NiMoO_4$, NiFe PBA/$NiMoO_4$ and $NiFe_2O_4$/$NiMoO_4$. Figure S4. XPS spectrum of (a) $NiMoO_4$, $NiMoO_4$ O_2-Pl, NiFe PBA/$NiMoO_4$, $NiFe_2O_4$/$NiMoO_4$, and of (b) $NiFe_2O_4$/$NiMoO_4$ before OER and after OER. Figure S5. XPS (a) Ni 2p and (b) Fe 2p spectra of NiFe PBA/$NiMoO_4$. Figure S6. XPS analysis of NiFe PBA/$NiMoO_4$ and $NiFe_2O_4$/$NiMoO_4$. The core level spectra of (a) Ni $2p_{3/2}$ and (b) Mo 3d. Figure S7. Cyclic voltammograms in a capacitive current region at various scan rates from 20 to 100 mV s^{-1}. (a) $NiMoO_4$, (b) $NiMoO_4$ O_2-Pl, (c) NiFe PBA/$NiMoO_4$, (d) $NiFe_2O_4$/$NiMoO_4$. Figure S8. In situ Raman spectra of $NiFe_2O_4$/$NiMoO_4$for activation from 1.18 V to 1.63 V (a) in a region from 250 cm^{-1} to 1050 cm^{-1} and (b) in a region from 1500 cm^{-1} to 2500 cm^{-1}. Table S1. Chemical composition of $NiFe_2O_4$/$NiMoO_4$ based on EDS.

Author Contributions: Conceptualization, N.X. and W.P.; methodology, N.X.; software, L.L., P.X., C.W. and J.L.; validation, W.P.; formal analysis, N.X. and W.P.; investigation, N.X.; resources, N.X.; data curation, N.X.; writing—original draft preparation, N.X.; writing—review and editing, N.X.; visualization, W.L. and L.Z.; supervision, W.L. and L.Z.; project administration, W.L. and L.Z.; funding acquisition, W.L. and L.Z. All authors contributed to the critical literature review. All authors contributed to writing and revising the manuscript. All authors have read and agreed to the published version of the manuscript.

Funding: This work was supported by the National Natural Science Foundation of China (51904216, 21905218), the Natural Science Foundation of Hubei Province (2019CFA001), the Foshan Xianhu Laboratory of the Advanced Energy Science and Technology Guangdong Laboratory (XHT2020-003), and the Fundamental Research Funds for the Central Universities (WUT: 2020IVB034, 2020IVA036, 2021CG014). The TEM work was performed at the Nanostructure Research Center (NRC), which was supported by the Fundamental Research Funds for the Central Universities (WUT: 2019III012GX), the State Key Laboratory of Advanced Technology for Materials Synthesis and Processing.

Institutional Review Board Statement: Not applicable.

Informed Consent Statement: Not applicable.

Data Availability Statement: The data presented in this study are available from the corresponding author, upon reasonable request.

Conflicts of Interest: The authors declare no conflict of interest.

References

1. Chu, S.; Majumdar, A. Opportunities and Challenges for a Sustainable Energy Future. *Nature* **2012**, *488*, 294–303. [CrossRef] [PubMed]
2. Jiao, Y.; Zheng, Y.; Jaroniec, M.; Qiao, S.Z. Design of Electrocatalysts for Oxygen- and Hydrogen-Involving Energy Conversion Reactions. *Chem. Soc. Rev.* **2015**, *44*, 2060–2086. [CrossRef] [PubMed]
3. Li, L.; Wang, P.; Shao, Q.; Huang, X. Recent Progress in Advanced Electrocatalyst Design for Acidic Oxygen Evolution Reaction. *Adv. Mater.* **2021**, *33*, 2004243. [CrossRef] [PubMed]
4. Suen, N.T.; Hung, S.F.; Quan, Q.; Zhang, N.; Xu, Y.J.; Chen, H.M. Electrocatalysis for the Oxygen Evolution Reaction: Recent Development and Future Perspectives. *Chem. Soc. Rev.* **2017**, *46*, 337–365. [CrossRef]
5. Gao, L.; Cui, X.; Sewell, C.D.; Li, J.; Lin, Z. Recent Advances in Activating Surface Reconstruction for the High-Efficiency Oxygen Evolution Reaction. *Chem. Soc. Rev.* **2021**, *50*, 8428–8469. [CrossRef]
6. Li, S.; Gao, Y.; Li, N.; Ge, L.; Bu, X.; Feng, P. Transition Metal-Based Bimetallic Mofs and Mof-Derived Catalysts for Electrochemical Oxygen Evolution Reaction. *Energy Environ. Sci.* **2021**, *14*, 1897–1927. [CrossRef]
7. Song, J.; Wei, C.; Huang, Z.F.; Liu, C.; Zeng, L.; Wang, X.; Xu, Z.J. A Review on Fundamentals for Designing Oxygen Evolution Electrocatalysts. *Chem. Soc. Rev.* **2020**, *49*, 2196–2214. [CrossRef]
8. Zou, X.; Zhang, Y. Noble Metal-Free Hydrogen Evolution Catalysts for Water Splitting. *Chem. Soc. Rev.* **2015**, *44*, 5148–5180. [CrossRef]
9. Yan, D.; Li, Y.; Huo, J.; Chen, R.; Dai, L.; Wang, S. Defect Chemistry of Nonprecious-Metal Electrocatalysts for Oxygen Reactions. *Adv. Mater.* **2017**, *29*, 1606459. [CrossRef]
10. Shi, Y.; Zhang, B. Recent Advances in Transition Metal Phosphide Nanomaterials: Synthesis and Applications in Hydrogen Evolution Reaction. *Chem. Soc. Rev.* **2016**, *45*, 1529–1541. [CrossRef]

11. Su, X.; Wang, Y.; Zhou, J.; Gu, S.; Li, J.; Zhang, S. Operando Spectroscopic Identification of Active Sites in Nife Prussian Blue Analogues as Electrocatalysts: Activation of Oxygen Atoms for Oxygen Evolution Reaction. *J. Am. Chem. Soc.* **2018**, *140*, 11286–11292. [CrossRef] [PubMed]
12. Gao, Q.; Huang, C.Q.; Ju, Y.M.; Gao, M.R.; Liu, J.W.; An, D.; Cui, C.H.; Zheng, Y.R.; Li, W.X.; Yu, S.H. Phase-Selective Syntheses of Cobalt Telluride Nanofleeces for Efficient Oxygen Evolution Catalysts. *Angew. Chem. Int. Ed.* **2017**, *56*, 7769–7773. [CrossRef] [PubMed]
13. Li, S.; Chen, B.; Wang, Y.; Ye, M.Y.; van Aken, P.A.; Cheng, C.; Thomas, A. Oxygen-Evolving Catalytic Atoms on Metal Carbides. *Nat. Mater.* **2021**, *20*, 1240–1247. [CrossRef] [PubMed]
14. Aralekallu, S.; Sajjan, V.A.; Palanna, M.; CP, K.P.; Hojamberdiev, M.; Sannegowda, L.K. Ni foam-supported azo linkage cobalt phthalocyanine as an efficient electrocatalyst for oxygen evolution reaction. *J. Power Sources* **2020**, *449*, 227516. [CrossRef]
15. Duan, Y.; Yu, Z.Y.; Hu, S.J.; Zheng, X.S.; Zhang, C.T.; Ding, H.H.; Hu, B.C.; Fu, Q.Q.; Yu, Z.L.; Zheng, X.; et al. Scaled-up Synthesis of Amorphous Nifemo Oxides and Their Rapid Surface Reconstruction for Superior Oxygen Evolution Catalysis. *Angew. Chem. Int. Ed. Engl.* **2019**, *58*, 15772–15777. [CrossRef]
16. Kuznetsov, D.A.; Han, B.; Yu, Y.; Rao, R.R.; Hwang, J.; Román-Leshkov, Y.; Shao-Horn, Y. Tuning Redox Transitions Via Inductive Effect in Metal Oxides and Complexes, and Implications in Oxygen Electrocatalysis. *Joule* **2018**, *2*, 225–244. [CrossRef]
17. Li, J.; Chu, D.; Dong, H.; Baker, D.R.; Jiang, R. Boosted Oxygen Evolution Reactivity by Igniting Double Exchange Interaction in Spinel Oxides. *J. Am. Chem. Soc.* **2020**, *142*, 50–54. [CrossRef]
18. Zhou, J.; Zhang, L.; Huang, Y.C.; Dong, C.L.; Lin, H.J.; Chen, C.T.; Tjeng, L.H.; Hu, Z. Voltage- and Time-Dependent Valence State Transition in Cobalt Oxide Catalysts During the Oxygen Evolution Reaction. *Nat. Commun.* **2020**, *11*, 1984. [CrossRef]
19. Zhou, D.; Li, P.; Lin, X.; McKinley, A.; Kuang, Y.; Liu, W.; Lin, W.F.; Sun, X.; Duan, X. Layered Double Hydroxide-Based Electrocatalysts for the Oxygen Evolution Reaction: Identification and Tailoring of Active Sites, and Superaerophobic Nanoarray Electrode Assembly. *Chem. Soc. Rev.* **2021**, *50*, 8790–8817. [CrossRef]
20. Gao, M.; Sheng, W.; Zhuang, Z.; Fang, Q.; Gu, S.; Jiang, J.; Yan, Y. Efficient Water Oxidation Using Nanostructured α-Nickel-Hydroxide as an Electrocatalyst. *J. Am. Chem. Soc.* **2014**, *136*, 7077–7084. [CrossRef]
21. Yang, H.; Gong, L.; Wang, H.; Dong, C.; Wang, J.; Qi, K.; Liu, H.; Guo, X.; Xia, B.Y. Preparation of Nickel-Iron Hydroxides by Microorganism Corrosion for Efficient Oxygen Evolution. *Nat. Commun.* **2020**, *11*, 5075. [CrossRef] [PubMed]
22. Govind Rajan, A.; Martirez, J.M.P.; Carter, E.A. Facet-Independent Oxygen Evolution Activity of Pure Beta-Niooh: Different Chemistries Leading to Similar Overpotentials. *J. Am. Chem. Soc.* **2020**, *142*, 3600–3612. [CrossRef] [PubMed]
23. Ye, S.H.; Shi, Z.X.; Feng, J.X.; Tong, Y.X.; Li, G.R. Activating CoOOH Porous Nanosheet Arrays by Partial Iron Substitution for Efficient Oxygen Evolution Reaction. *Angew. Chem. Int. Ed.* **2018**, *57*, 2672–2676. [CrossRef] [PubMed]
24. Han, X.; Yu, C.; Zhou, S.; Zhao, C.; Huang, H.; Yang, J.; Liu, Z.; Zhao, J.; Qiu, J. Ultrasensitive Iron-Triggered Nanosized Fe-CoOOH Integrated with Graphene for Highly Efficient Oxygen Evolution. *Adv. Energy Mater.* **2017**, *7*, 1602148. [CrossRef]
25. Huang, Z.-F.; Song, J.; Du, Y.; Xi, S.; Dou, S.; Nsanzimana, J.M.V.; Wang, C.; Xu, Z.J.; Wang, X. Chemical and Structural Origin of Lattice Oxygen Oxidation in Co-Zn Oxyhydroxide Oxygen Evolution Electrocatalysts. *Nat. Energy* **2019**, *4*, 329–338. [CrossRef]
26. Sun, H.; Yan, Z.; Liu, F.; Xu, W.; Cheng, F.; Chen, J. Self-Supported Transition-Metal-Based Electrocatalysts for Hydrogen and Oxygen Evolution. *Adv. Mater.* **2020**, *32*, 1806326. [CrossRef]
27. Peugeot, A.; Creissen, C.E.; Karapinar, D.; Tran, H.N.; Schreiber, M.; Fontecave, M. Benchmarking of Oxygen Evolution Catalysts on Porous Nickel Supports. *Joule* **2021**, *5*, 1281–1300. [CrossRef]
28. Liu, Q.; Zhao, H.; Jiang, M.; Kang, Q.; Zhou, W.; Wang, P.; Zhou, F. Boron Enhances Oxygen Evolution Reaction Activity over Ni Foam-Supported Iron Boride Nanowires. *J. Mater. Chem. A* **2020**, *8*, 13638–13645. [CrossRef]
29. Li, R.-Q.; Wan, X.-Y.; Chen, B.-L.; Cao, R.-Y.; Ji, Q.-H.; Deng, J.; Qu, K.-G.; Wang, X.-B.; Zhu, Y.-C. Hierarchical $Ni_3N/Ni_{0.2}Mo_{0.8}N$ Heterostructure Nanorods Arrays as Efficient Electrocatalysts for Overall Water and Urea Electrolysis. *Chem. Eng. J.* **2021**, *409*, 128240. [CrossRef]
30. Wang, C.; Tian, Y.; Gu, Y.; Xue, K.-H.; Sun, H.; Miao, X.; Dai, L. Plasma-Induced Moieties Impart Super-Efficient Activity to Hydrogen Evolution Electrocatalysts. *Nano Energy* **2021**, *85*, 106030. [CrossRef]
31. Yin, Z.; Sun, Y.; Zhu, C.; Li, C.; Zhang, X.; Chen, Y. Bimetallic Ni-Mo Nitride Nanotubes as Highly Active and Stable Bifunctional Electrocatalysts for Full Water Splitting. *J. Mater. Chem. A* **2017**, *5*, 13648–13658. [CrossRef]
32. Durr, R.N.; Maltoni, P.; Tian, H.; Jousselme, B.; Hammarstrom, L.; Edvinsson, T. From $NiMoO_4$ to γ-NiOOH: Detecting the Active Catalyst Phase by Time Resolved in Situ and Operando Raman Spectroscopy. *ACS Nano* **2021**, *15*, 13504–13515. [CrossRef] [PubMed]
33. Wang, J.; Cui, W.; Liu, Q.; Xing, Z.; Asiri, A.M.; Sun, X. Recent Progress in Cobalt-Based Heterogeneous Catalysts for Electrochemical Water Splitting. *Adv. Mater.* **2016**, *28*, 215–230. [CrossRef] [PubMed]
34. Suryanto, B.H.R.; Wang, Y.; Hocking, R.K.; Adamson, W.; Zhao, C. Overall Electrochemical Splitting of Water at the Heterogeneous Interface of Nickel and Iron Oxide. *Nat. Commun.* **2019**, *10*, 5599. [CrossRef]
35. Guo, Y.; Yuan, P.; Zhang, J.; Xia, H.; Cheng, F.; Zhou, M.; Li, J.; Qiao, Y.; Mu, S.; Xu, Q. Co_2P-CoN Double Active Centers Confined in N-Doped Carbon Nanotube: Heterostructural Engineering for Trifunctional Catalysis toward Her, ORR, OER, and Zn-Air Batteries Driven Water Splitting. *Adv. Funct. Mater.* **2018**, *28*, 1805641. [CrossRef]

36. Wu, A.; Gu, Y.; Xie, Y.; Tian, C.; Yan, H.; Wang, D.; Zhang, X.; Cai, Z.; Fu, H. Effective Electrocatalytic Hydrogen Evolution in Neutral Medium Based on 2d MoP/MoS_2 Heterostructure Nanosheets. *ACS Appl. Mater. Interfaces* **2019**, *11*, 25986–25995. [CrossRef]
37. Lu, K.; Liu, Y.; Lin, F.; Cordova, I.A.; Gao, S.; Li, B.; Peng, B.; Xu, H.; Kaelin, J.; Coliz, D.; et al. Li_xNiO/Ni Heterostructure with Strong Basic Lattice Oxygen Enables Electrocatalytic Hydrogen Evolution with Pt-Like Activity. *J. Am. Chem. Soc.* **2020**, *142*, 12613–12619. [CrossRef]
38. Sun, J.; Xu, W.; Lv, C.; Zhang, L.; Shakouri, M.; Peng, Y.; Wang, Q.; Yang, X.; Yuan, D.; Huang, M.; et al. Co/MoN Hetero-Interface Nanoflake Array with Enhanced Water Dissociation Capability Achieves the Pt-Like Hydrogen Evolution Catalytic Performance. *Appl. Catal. B* **2021**, *286*, 119882. [CrossRef]
39. Lv, J.; Wang, L.; Li, R.; Zhang, K.; Zhao, D.; Li, Y.; Li, X.; Huang, X.; Wang, G. Constructing a Hetero-Interface Composed of Oxygen Vacancy-Enriched Co_3O_4 and Crystalline-Amorphous Nife-LDH for Oxygen Evolution Reaction. *ACS Catal.* **2021**, *11*, 14338–14351. [CrossRef]
40. Zhang, W.; Wang, Y.; Zheng, H.; Li, R.; Tang, Y.; Li, B.; Zhu, C.; You, L.; Gao, M.R.; Liu, Z.; et al. Embedding Ultrafine Metal Oxide Nanoparticles in Monolayered Metal-Organic Framework Nanosheets Enables Efficient Electrocatalytic Oxygen Evolution. *ACS Nano* **2020**, *14*, 1971–1981. [CrossRef]
41. CP, K.P.; Aralekallu, S.; Sajjan, V.A.; Palanna, M.; Kumar, S.; Sannegowda, L.K. Non-precious cobalt phthalocyanine-embedded iron ore electrocatalysts for hydrogen evolution reactions. *Sustain. Energy Fuels* **2021**, *5*, 1448–1457. [CrossRef]
42. Bai, L.; Lee, S.; Hu, X. Spectroscopic and Electrokinetic Evidence for a Bifunctional Mechanism of the Oxygen Evolution Reaction. *Angew. Chem. Int. Ed.* **2021**, *60*, 3095–3103. [CrossRef] [PubMed]
43. Clament Sagaya Selvam, N.; Kwak, S.J.; Choi, G.H.; Oh, M.J.; Kim, H.; Yoon, W.-S.; Lee, W.B.; Yoo, P.J. Unveiling the Impact of Fe Incorporation on Intrinsic Performance of Reconstructed Water Oxidation Electrocatalyst. *ACS Energy Lett.* **2021**, *6*, 4345–4354. [CrossRef]
44. Zhong, L.; Bao, Y.; Feng, L. Fe-Doping Effect on CoTe Catalyst with Greatly Boosted Intrinsic Activity for Electrochemical Oxygen Evolution Reaction. *Electrochim. Acta* **2019**, *321*, 134656. [CrossRef]
45. Guo, D.; Luo, Y.; Yu, X.; Li, Q.; Wang, T. High Performance $NiMoO_4$ Nanowires Supported on Carbon Cloth as Advanced Electrodes for Symmetric Supercapacitors. *Nano Energy* **2014**, *8*, 174–182. [CrossRef]
46. Choi, J.; Kim, D.; Zheng, W.; Yan, B.; Li, Y.; Lee, L.Y.S.; Piao, Y. Interface Engineered $NiFe_2O_{4-x}$/$NiMoO_4$ Nanowire Arrays for Electrochemical Oxygen Evolution. *Appl. Catal. B* **2021**, *286*, 119857. [CrossRef]
47. Xie, Y.; Chen, C.; Lu, X.; Luo, F.; Wang, C.; Alsaedi, A.; Hayat, T. Porous NiFe-Oxide Nanocubes Derived from Prussian Blue Analogue as Efficient Adsorbents for the Removal of Toxic Metal Ions and Organic Dyes. *J. Hazard. Mater.* **2019**, *379*, 120786. [CrossRef]
48. Wang, J.; Sun, M.; Wang, F.; Zhang, X.; Song, J.; Zhang, W.; Li, M. Essential Role of the Interfacial Interaction in the Core-Shell Electrode for Its Enhanced Electrochemical Performance. *Chem. Eng. J.* **2021**, *426*, 130895. [CrossRef]
49. Zhang, W.; Zhao, Y.; Malgras, V.; Ji, Q.; Jiang, D.; Qi, R.; Ariga, K.; Yamauchi, Y.; Liu, J.; Jiang, J.S.; et al. Synthesis of Monocrystalline Nanoframes of Prussian Blue Analogues by Controlled Preferential Etching. *Angew. Chem. Int. Ed.* **2016**, *55*, 8228–8234. [CrossRef]
50. Yilmaz, G.; Tan, C.F.; Lim, Y.-F.; Ho, G.W. Pseudomorphic Transformation of Interpenetrated Prussian Blue Analogs into Defective Nickel Iron Selenides for Enhanced Electrochemical and Photo-Electrochemical Water Splitting. *Adv. Energy Mater.* **2019**, *9*, 1802983. [CrossRef]
51. Ahmed, Z.; Krishankant; Rai, R.; Kumar, R.; Maruyama, T.; Bera, C.; Bagchi, V. Unraveling a Graphene Exfoliation Technique Analogy in the Making of Ultrathin Nickel-Iron Oxyhydroxides@Nickel Foam to Promote the OER. *ACS Appl. Mater. Interfaces* **2021**, *13*, 55281–55291. [CrossRef] [PubMed]
52. Das, D.; Santra, S.; Nanda, K.K. In Situ Fabrication of a Nickel/Molybdenum Carbide-Anchored N-Doped Graphene/Cnt Hybrid: An Efficient (Pre)Catalyst for OER and HER. *ACS Appl. Mater. Interfaces* **2018**, *10*, 35025–35038. [CrossRef] [PubMed]
53. Wu, Z.; Zou, Z.; Huang, J.; Gao, F. $NiFe_2O_4$ Nanoparticles/Nife Layered Double-Hydroxide Nanosheet Heterostructure Array for Efficient Overall Water Splitting at Large Current Densities. *ACS Appl. Mater. Interfaces* **2018**, *10*, 26283–26292. [CrossRef] [PubMed]
54. Chung, D.Y.; Lopes, P.P.; Farinazzo Bergamo Dias Martins, P.; He, H.; Kawaguchi, T.; Zapol, P.; You, H.; Tripkovic, D.; Strmcnik, D.; Zhu, Y.; et al. Dynamic Stability of Active Sites in Hydr(Oxy)Oxides for the Oxygen Evolution Reaction. *Nat. Energy* **2020**, *5*, 222–230. [CrossRef]
55. Zhang, B.; Wu, Z.; Shao, W.; Gao, Y.; Wang, W.; Ma, T.; Ma, L.; Li, S.; Cheng, C.; Zhao, C. Interfacial Atom-Substitution Engineered Transition-Metal Hydroxide Nanofibers with High-Valence Fe for Efficient Electrochemical Water Oxidation. *Angew. Chem. Int. Ed.* **2022**, *61*, 202115331.
56. Zhang, B.; Wang, L.; Cao, Z.; Kozlov, S.M.; García de Arquer, F.P.; Dinh, C.T.; Li, J.; Wang, Z.; Zheng, X.; Zhang, L.; et al. High-Valence Metals Improve Oxygen Evolution Reaction Performance by Modulating 3d Metal Oxidation Cycle Energetics. *Nat. Catal.* **2020**, *3*, 985–992. [CrossRef]
57. Lin, Y.C.; Chuang, C.H.; Hsiao, L.Y.; Yeh, M.H.; Ho, K.C. Oxygen Plasma Activation of Carbon Nanotubes-Interconnected Prussian Blue Analogue for Oxygen Evolution Reaction. *ACS Appl. Mater. Interfaces* **2020**, *12*, 42634–42643. [CrossRef]

58. Nieuwoudt, M.K.; Comins, J.D.; Cukrowski, I. The Growth of the Passive Film on Iron in 0.05 M NaOH Studied in Situ by Rama Micro-Spectroscopy and Electrochemical Polarisation. Part I: Near-Resonance Enhancement of the Raman Spectra of Iron Oxic and Oxyhydroxide Compounds. *J. Raman Spectrosc.* **2011**, *42*, 1335–1339. [CrossRef]
59. Lee, S.-Y.; Kim, I.-S.; Cho, H.-S.; Kim, C.-H.; Lee, Y.-K. Resolving Potential-Dependent Degradation of Electrodeposited Ni(OH Catalysts in Alkaline Oxygen Evolution Reaction (OER): In Situ XANES Studies. *Appl. Catal. B* **2021**, *284*, 119729. [CrossRef]
60. Mansour, A.; Melendres, C.J.S.S.S. Characterization of Electrochemically Prepared γ-NiOOH by XPS. *Surf. Sci. Spectra* **1994**, 271–278. [CrossRef]
61. Tan, B.J.; Klabunde, K.J.; Sherwood, P.M. X-ray Photoelectron Spectroscopy Studies of Solvated Metal Atom Dispersed Catalys Monometallic Iron and Bimetallic Iron-Cobalt Particles on Alumina. *Chem. Mater.* **1990**, *2*, 186–191. [CrossRef]
62. Welsh, I.D.; Sherwood, P.M.A. Photoemission and Electronic Structure of FeOOH: Distinguishing between Oxide and Oxyhydro ide. *Phys. Rev. B* **1989**, *40*, 6386. [CrossRef] [PubMed]

materials

MDPI

Article

Photoelectric Properties of Planar and Mesoporous Structured Perovskite Solar Cells

Steponas Ašmontas *, Aurimas Čerškus, Jonas Gradauskas, Asta Grigucevičienė, Remigijus Juškėnas, Konstantinas Leinartas, Andžej Lučun, Kazimieras Petrauskas, Algirdas Selskis, Laurynas Staišiūnas, Algirdas Sužiedėlis, Aldis Šilėnas and Edmundas Širmulis

Center for Physical Sciences and Technology, Savanorių Ave. 231, LT-02300 Vilnius, Lithuania; aurimas.cerskus@ftmc.lt (A.Č.); jonas.gradauskas@ftmc.lt (J.G.); asta.griguceviciene@ftmc.lt (A.G.); remigijus.juskenas@ftmc.lt (R.J.); konstantinas.leinartas@ftmc.lt (K.L.); andzej.lucun@ftmc.lt (A.L.); kazimieras.petrauskas@ftmc.lt (K.P.); algirdas.selskis@ftmc.lt (A.S.); laurynas.staisiunas@ftmc.lt (L.S.); algirdas.suziedelis@ftmc.lt (A.S.); aldis.silenas@ftmc.lt (A.Š.); edmundas.sirmulis@ftmc.lt (E.Š.)

* Correspondence: steponas.asmontas@ftmc.lt; Tel.: +370-5-2627-124

Abstract: The high efficiency of perovskite solar cells strongly depends on the quality of perovskite films and carrier extraction layers. Here, we present the results of an investigation of the photoelectric properties of solar cells based on perovskite films grown on compact and mesoporous titanium dioxide layers. Kinetics of charge carrier transport and their extraction in triple-cation perovskite solar cells were studied by using transient photovoltage and time-resolved photoluminescence decay measurements. X-ray diffraction analysis revealed that the crystallinity of the perovskite films grown on mesoporous titanium dioxide is better compared to the films grown on compact TiO_2. Mesoporous structured perovskite solar cells are found to have higher power conversion efficiency mainly due to enlarged perovskite/mesoporous -TiO_2 interfacial area and better crystallinity of their perovskite films.

Keywords: advanced energy materials; perovskite; solar cell; thin film; power conversion efficiency

1. Introduction

It has become increasingly urgent to look for renewable energy sources and improve the existing ones due to the ever-increasing energy demand accompanied by the exhaustion of organic fuels, global warming, and the current political situation. One of the most promising and environmentally friendly energy sources is electricity generated by solar cells. The organic–inorganic hybrid perovskite-based solar cells (PSC) show a promising future for photovoltaic technology [1]. During the last decade, the power conversion efficiency (PCE) of PSCs has grown from 3.8% to more than 25.7% [2], making them a very rapidly advancing technology and a popular topic in the field of solar cells. The high performance of PSCs can primarily be attributed to their excellent optoelectronic properties, such as (1) strong light absorption through the entire visible spectrum range, enabling the use of thin perovskite films [3,4]; (2) high carrier mobility [5,6]; (3) long carrier diffusion length [7,8]; (4) high defect tolerance [9]; and (5) long lifetime of the generated charge carriers [10,11]. To realize the good performance of a PSC, high-quality perovskite film containing a low density of defects must be employed in a single device [12]. The defects in a perovskite layer may affect both charge carrier transport and their extraction processes [13], and they can induce trap states, causing increased charge carrier recombination rate and, thereby, the reduced power conversion efficiency of a PSC [14]. To obtain perfectly operating perovskite devices, carrier transport and extraction layers are also important in addition to high-quality perovskite films [15,16]. There are several basic requirements for the transport layers: (1) good energy-level alignment with a perovskite for efficient charge transfer, (2) high mobility of charge carriers to guarantee their fast transport, (3) high transmittance to reduce the optical energy loss, (4) good stability, and (5) easy processing and low cost [16]. Two typical PSC structures are widely used: one is a planar heterojunction architecture [7,17], and another is a mesoporous structure [18,19]. Planar perovskite solar cells are advantageous because they have simple and scalable configurations [20]. However, the highest PCE is achieved with mesoscopic PSC, in which a mesoporous metal oxide layer plays an important role as the charge transport channel, scaffold for loading the light-absorbing materials, and electron–hole separator [21].

Titanium dioxide (TiO_2) has been mostly used as an electron transport layer (ETL) in PSCs due to its wide bandgap, E_g = 3.2 eV, suitable for minimized parasitic absorption, an appropriate conduction band alignment with a perovskite for efficient electron injection, and good electron-transport capability [22]. The quality of ETL plays an important role in improving the performance of PSCs [23]. An ETL transports photoinduced electrons away from the perovskite, and at the same time, it serves as a blocking layer preventing direct contact between the holes and fluorine-doped tin oxide (FTO) [24,25]. Therefore, TiO_2 layers should be uniform, pinhole-free, and should completely cover the surface of FTO [26]. The optimal thickness of crystalline TiO_2 depends on the deposition technique and post-treatment of the layers. In the case of planar PSCs, the thickness of an ETL is in the range of 20–30 nm [26–28]. The deposition of a mesoporous (mp-TiO_2) layer on crystalline titanium dioxide provides a reduced carrier transport length and increased interfacial contact area, thereby facilitating efficient charge carrier collection and yielding the highest reported PCE values [29]. It has been shown that the structural properties of the mp-TiO_2 layer, such as particle size, thickness, and porosity, have a significant influence not only on the performance of PSCs but also on the layer's photoelectric properties [13,26,30–32]. The transient photovoltage (TPV) measurements showed that the TPV decay strongly depends on the porosity of TiO_2 film, and there is a significant correlation between the TPV decay characteristics and the performance of the perovskite solar cells [13]. It was established that the time-resolved photoluminescence (TRPL) decay also strongly depends on the size of TiO_2 nanoparticles [30], the porosity of the TiO_2 film, mesoscopic structure, and perovskite morphology formed therein [13]. TRPL spectroscopy is a powerful technique to study excited charge carrier recombination in perovskite layers that notably determine the PCE of PSCs [14].

In this paper, we report experimental results of the kinetics of the charge carrier transport and their extraction obtained by performing transient photovoltage and time-resolved photoluminescence decay measurements with the intention to elucidate the correlation between the electronic processes in perovskite layers and the performance of perovskite solar cells.

2. Fabrication of Perovskite Films and Their Characterization

In this study, two types of perovskite solar cells, with and without a porous TiO_2 layer, were fabricated. Their schematic configurations are shown in Figure 1. The cells also differ in the formation and composition of the compact TiO_2 layer: in one case, the compact TiO_2 layers were formed by the conventional spray pyrolysis method; in another case, the compact TiO_2 layers containing niobium oxide ($Ti_{1-x}Nb_xO_{2-}$ mixed titanium niobium oxide) were deposited using the Atomic Layer Deposition (ALD) method. In both cases, the 25 × 25 mm^2 glasses coated with a transparent fluorine-doped tin oxide (FTO) layer (TEC 10, Ossila, UK) were used as substrates. The composition of precursors, their concentrations and purity grades, the sequence of procedures and equipment, and the parameters of the layers (ETL, perovskite layer, and hole transport layer), all used for the fabrication of the perovskite cells, are detailed in our previous papers [33,34].

Thin, compact $Ti_{0.93}Nb_{0.07}O_2$ layers were deposited in a "Fiji F200" ALD reactor (Cambridge Nano Tech, Cambridge, MA, USA) using similar procedures as described in [35]. Precursors for titanium and niobium oxides were tetrakis dimethylamino titanium (TDMAT, 99.9%; STREM Chemicals Inc., Newburyport, MA, USA) and niobium ethoxide ($Nb(OEt)_5$, 99.9%; STREM Chemicals Inc., Newburyport, MA, USA), and deionized water was used as an oxygen source. The mixed oxide was formed by depositing a single layer of niobium oxide for every 20 layers of deposited titanium oxide. The deposition consisted of 420 monolayers and resulted in an approximately 30 nm-thick coating.

(a) (b)

Figure 1. Schematic configuration of the cells fabricated for the study: (**a**) without porous TiO_2 layer, (**b**) with porous TiO_2 layer. The compact TiO_2 layers with and without Nb_2O_5 were formed using the ALD and spray pyrolysis methods, respectively.

We should note that porous TiO_2, perovskite, and hole transport layers were formed over the ALD-mixed titanium niobium oxide layer using the same precursors and procedures as in the case of compact TiO_2 layers formed by spray pyrolysis. Cs-containing $Cs_x(MA_{0.17}FA_{0.83})_{1-x}Pb(I_{0.83}Br_{0.17})_3$ perovskite layers of the same composition with x = 0.1 were grown for both cell types. The 70 nm-thick Au contacts were deposited in the vacuum chamber of the thermal evaporation equipment "VAKSIS PVD Vapor-5S_Th" (Vaksis R&D and Engineering, Ankara, Turkey) on the top of the hole transport layer (Spiro-OMeTAD from Sigma-Aldrich, St. Louis, MO, USA, was used for its formation) and on the open side of FTO (see Figure 1).

The surface morphology and cross-section of the films were examined using a scanning electron microscope (SEM) (Helios NanoLab 650, FEI, Hillsboro, OR, USA).

X-ray diffraction (XRD) patterns were obtained using an X-ray diffractometer (XRD) (SmartLab, Rigaku, Tokyo, Japan) equipped with a 9 kW power rotating Cu anode X-ray source and theta/theta goniometer. The patterns were measured with Bragg–Brentano geometry in the 2Θ range of 10–65°.

The transient photovoltage of the solar cells was measured using the setup shown in Figure 2. A diode-pumped, frequency-doubled Nd:YAG-LBO laser NL202 (Ekspla Ltd., Vilnius, Lithuania) generated 7 ns-long pulses of 532 nm radiation. To prevent the sample from heating, a low pulse repetition rate of 5 Hz was used. A harmonic separator (dichroic beamsplitter; Eksma Ltd., Vilnius, Lithuania) was employed to reflect the rest of the infrared beam and transmit the 532 nm radiation. A part of the laser beam was directed to an optical signal reference detector (Standa Ltd., Vilnius, Lithuania) using a beam splitter plate. A variable attenuator (Eksma Ltd., Vilnius, Lithuania) consisting of a Brewster-type polarizer and quartz half wave-plate allowed changing the power of the incident laser beam. The average power of the laser radiation was measured by an optical power meter PM400 (Thorlabs Inc., Newton, NJ, USA). The maximum power density incident on the sample was about 0.5 $mWcm^{-2}$. A laser beam was directed onto the sample S with a removable high-reflectance mirror (M). The transient photovoltage and the laser pulse were recorded using a digital storage oscilloscope DSO6102A (Agilent Technologies Inc., Santa Clara, CA, USA), and the laser pulse shape was registered by the high-speed detector 11HSP-FS1 (Standa Ltd., Vilnius, Lithuania).

The light from a continuous-wave (CW) Ar-ion laser or pulsed 532 nm microchip laser (Standa Ltd., Vilnius, Lithuania) was used for excitation in a standard photoluminescence (PL) measurement setup with a 1 m monochromator and thermoelectrically cooled photomultiplier tube working in the photon counting regime. More details of this technique are described in our previous paper [36]. The power of the lasers was varied with a linear variable metallic neutral density filter. The samples were held in a vacuumed optical cryostat during all CWPL and time-resolved photoluminescence (TRPL) measurements at room temperature.

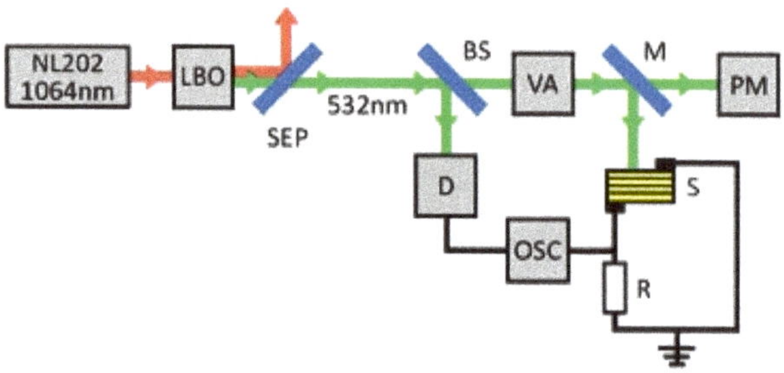

Figure 2. The scheme of the experimental setup of transient photovoltage measurement. NL202 is the diode-pumped Nd:YAG laser; LBO is the nonlinear crystal for the second harmonic generation; SEP is the Nd:YAG laser harmonic separator; BS is the beam splitter plate; VA is the variable attenuator for 532 nm laser beam; M is the removable high-reflectance mirror; S is the sample solar cell; R is the load resistor; PM is the optical power meter; D is the optical signal reference detector; OSC is the digital storage oscilloscope.

The photovoltaic characteristics of the perovskite solar cells were measured using Keithley 2602A (Keithley Instruments Inc., Cleveland, OH, USA) equipment. The 100 mW/cm^2 irradiance was produced by an AM 1.5 spectral lamp (Newport model 67005, Newport Corp., Irvine, CA, USA) placed at an appropriate distance.

3. Results and Discussion

Figure 3 presents the XRD patterns of the $Ti_{0.93}Nb_{0.07}O_2$ film before (as-deposited) and after annealing in an H_2 atmosphere at 400 °C for 50 min. The ramping-up rate was 5 °C/min, and the ramping-down rate was 3 °C per minute. It can be seen that the XRD pattern of the as-deposited film presents only FTO peaks. The XRD peaks of TiO_2 (anatase) emerged on the XRD pattern of the $Ti_{0.93}Nb_{0.07}O_2$ film annealed in a H_2 atmosphere. The electrical resistivity of the $Ti_{0.93}Nb_{0.07}O_2$ film decreased from 2×10^2 to 3.5×10^{-3} Ω cm during the annealing.

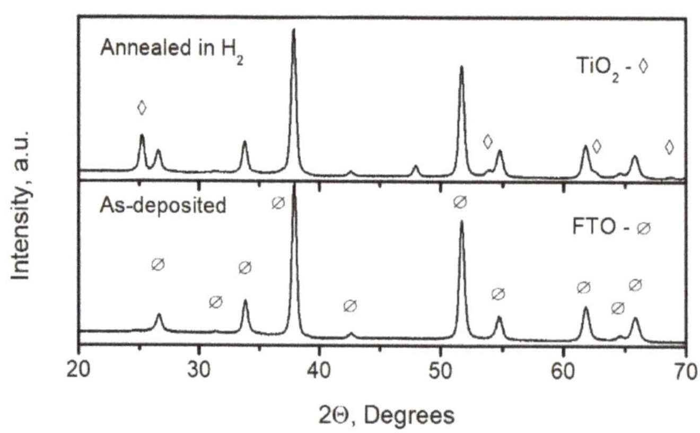

Figure 3. XRD patterns of the $Ti_{0.93}Nb_{0.07}O_2$ film before (as-deposited) and after annealing in a H_2 atmosphere.

The top-view SEM images of the triple cation perovskite films grown on titanium dioxide using different methods are shown in Figure 4. The first perovskite (Film A) was grown on a compact titanium dioxide (c-TiO_2) layer formed by the conventional spray pyrolysis method (FTO/c-TiO_2/perovskite). The second perovskite film (Film B) was grown on the mesoporous TiO_2 layer (FTO/c-TiO_2/mp-TiO_2/perovskite). The third perovskite film (Film C) was grown on the compact TiO_2 layer deposited using the ALD method (FTO/ALD-TiO_2/perovskite). The fourth perovskite film (Film D) was grown on the mesoporous TiO_2 layer (FTO/ALD-TiO_2/mp-TiO_2/perovskite). It can be seen that the morphology of the perovskite films and grain sizes are similar and the surfaces are flat and smooth, in agreement with other reports elsewhere [23,32].

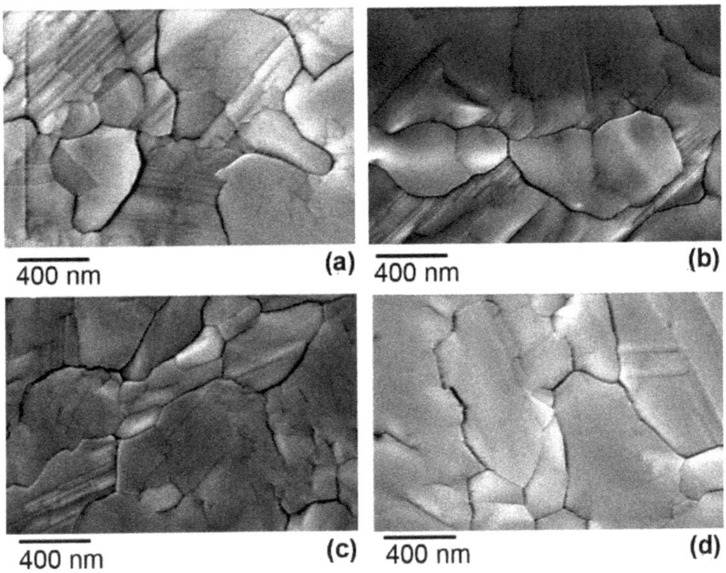

Figure 4. Surface SEM images of perovskite films grown on different types of titanium dioxide layers: (**a**) Film A on the c-TiO_2 layer, (**b**) Film B on the mp-TiO_2 layer/c-TiO_2, (**c**) Film C on the compact ALD-TiO_2 layer, and (**d**) Film D on the mp-TiO_2 layer/ALD-TiO_2.

The cross-sectional SEM images of the same perovskite films are shown in Figure 5. It is seen that the morphology and grain distribution in the films grown on the compact TiO_2 are similar (Figure 5a,c). The thickness of Films A and C was (880 $\pm$ 30) nm and (1200 $\pm$ 50) nm, respectively. However, the interface between Film A and c-TiO_2 formed by spray pyrolysis (Figure 5a) contained some small cavities. The perovskite material grown on mesoporous TiO_2 has penetrated between the nanocrystals of the mp-TiO_2 layer (Figure 5b,d). The thickness of Films B and D was (880 $\pm$ 50) nm and (720 $\pm$ 50) nm, respectively. Though the surface images of the perovskite films (Figure 4) show grains with dimensions greater than 1000 nm, the cross-section images show that the films were composed of stacked grains but not of ones spanning over the whole thickness of the layer (Figure 5b,d).

The XRD patterns of the perovskite films grown on different TiO_2 layers are depicted in Figure 6. Most of the XRD peaks corresponded to 2Θ values of the cubic $FA_{5/6}Ma_{1/6}PbBr_{0.5}I_{2.5}$ phase (ICDD PDF card #01-085-6374). The rest of the peaks were attributable to FTO. Visually, the XRD patterns of the films were very similar. However, in the case of Film C, the perovskite XRD peak 100 was of higher intensity than peak 110, in contrast to the other patterns. This could mean that the crystallites of the Film C have a prevailing orientation in the [100] direction while the rest of the films have another prevailing orientation, namely, [110].

The XRD peaks of the perovskite films formed on TiO_2 by spray pyrolysis seem to be sharper. This means that the crystallinity of the perovskite films grown on mesoporous titanium dioxide is better compared to the films grown on compact TiO_2. The values of full width at half-maximum (FWHM) of the perovskite peaks 100, 110, and 220 are presented in Table 1. The size of a crystallite, D, and the value of micro-deformation, $\Delta d/d$ (d is the interplanar spacing), both calculated using the Hall method, are also presented in the table. The FWHM values of Films A and B were smaller in comparison to those formed using the ALD method (Films C and D). Respectively, the size of crystallites was larger, and microdeformation was smaller in Films A and B, thus pointing to lower dislocation density in these films.

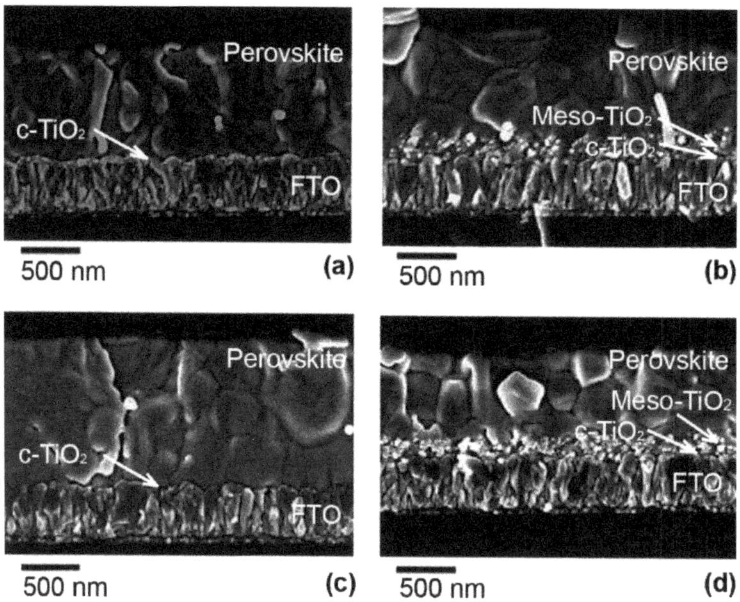

Figure 5. Cross-sectional SEM images of perovskite films on different TiO_2 layers: (**a**) Film A on the compact TiO_2 layer, (**b**) Film B on the mp-TiO_2 layer/c-TiO_2, (**c**), Film C on the compact ALD-TiO_2 layer, and (**d**) Film D on the mp-TiO_2 layer/ALD-TiO_2.

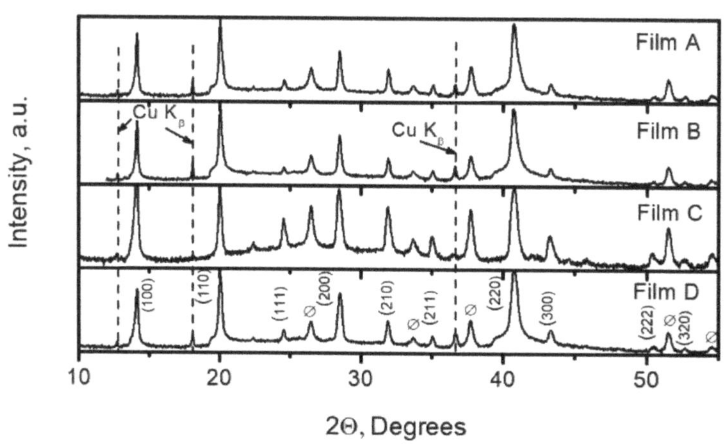

Figure 6. XRD patterns of perovskite films on different titanium dioxide layers: Film A is on the compact TiO_2 layer, Film B is on the mp-TiO_2 layer/c, Film C is on the compact ALD-TiO_2 layer, and Film D is on the ALD-TiO_2/mp-TiO_2 layer.

Table 1. The FWHM values of perovskite XRD peaks 100, 110, and 220, crystallite size D, and the value of microdeformation Δd/d in different perovskite films.

Sample	$FWHM_{(100)}$	$FWHM_{(110)}$	$FWHM_{(220)}$	D, nm	Δd/d
Film A	0.071	0.091	0.167	181.0 ± 30.3	0.25 ± 0.12
Film B	0.065	0.083	0.164	189.0 ± 82.3	0.23 ± 0.05
Film C	0.131	0.197	0.338	73.7 ± 15.0	0.29 ± 0.04
Film D	0.105	0.134	0.243	100.0 ± 12.3	0.30 ± 0.17

Figure 7 shows CW photoluminescence spectra measured from both sides of the planar and mesoporous structured solar cells. For most of the samples, the peak position of the top side (spiro-OMeTAD side) spectrum is within the ±3 nm range centered at 762 nm. This finding agrees well with the other results of triple cation perovskites containing 10% of Cs [34,37]. However, the 10 nm blue shift is detected in the case of the Film C sample. Nevertheless, the peak position of the bottom side (glass side) is less scattered among the samples (Figure 7b), but it is slightly redshifted compared to the spiro-OMeTAD side case. Analyzing the "same side" PL results, one can observe that the peak position of the samples containing the mp-TiO_2 layer has lower energy values than those of the samples with the c-TiO_2 layer; but, again, except for the sample with Film C from the spiro-OMeTAD side. A study of the integrated PL intensity I_{PL} dependence on laser power density P in the 1 mW/cm^2 to 65 mW/cm^2 range revealed that the Film C sample has the weakest dependence: its slope coefficient k ($\log I_{PL} \propto k \log P$) equals 0.95 when measured from the spiro-OMeTAD side. The coefficient k values are presented in Table 2. As a rule, the slope is below 1 for the free-to-bound and donor–acceptor transitions, and stays between 1 and 2 for the free-exciton or bound-exciton transitions [38]. The magnitude of k depends on the power density P and on material properties such as radiative and competitive nonradiative recombination probabilities [39] or the contribution of deep defects [40]. All these anomalies of Film C could be related to the smaller size of its crystallites. Therefore, a blue shift is observed in the CWPL spectrum from the spiro-OMeTAD side, and it changes the recombination probabilities, leading to a lower k value. Additionally, the decrease in the k value could result from donor–acceptor or free-bound transitions, including the contribution of the deep defects.

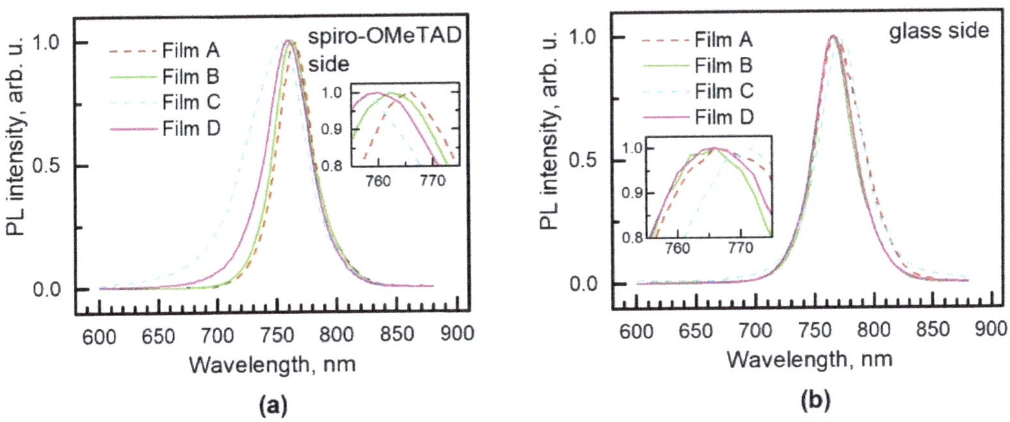

Figure 7. CWPL spectra of PSC measured from (**a**) Spiro-OMeTAD side and (**b**) Glass side. Dashed and solid lines indicate planar and mesoporous structured cells, respectively. The insets show the corresponding enlarged part of peak position.

Table 2. Results of PL spectra analysis. k is the slope of log I_{PL} versus log P curve; A_i and τ_i are the amplitude and the decay time constant of the ith term in Equation (1); τ_{dec} is average decay time calculated using Equation (2).

	Spiro-OMeTAD Side				Glass Side			
	Film A	Film B	Film C	Film D	Film A	Film B	Film C	Film D
k	1.37	1.57	0.95	1.22	1.36	1.42	1.11	1.49
A_1	0.85	0.82	0.40	0.78	0.64	0.49	0.50	0.69
τ_1, ns	2.5	2.9	0.79	2.2	3.2	2.7	0.96	1.4
A_2	0.19	0.16	0.37	0.23	0.37	0.38	0.44	0.29
τ_2, ns	23.1	27.4	3.3	10.4	15.6	27.7	4.7	9.5
A_3	0.05	0.05	0.08	0.05	0.04	0.11	0.07	0.07
τ_3, ns	95.9	124	15.8	54.4	105	119	28.2	45.9
τ_{dec}, ns	48.6	70.7	8.6	25.7	46.3	74.1	14.2	24.9

Time-resolved photoluminescence decays measured both from spiro-OMeTAD and glass sides under a pulsed laser fluence of 0.1 μJ/cm^2 are displayed in Figure 8. It shows the results of the longest average decay times obtained at different positions of all 20 samples (five samples from each series were tested). The transients were fitted using triple exponential approximation [33,41,42],

$$I_{PL}(t) = \sum_{i=1}^{3} A_i e^{-\frac{t}{\tau_i}} \quad (1)$$

where fitting parameters A_i and τ_i are the amplitude and the decay time constant of the ith term. The fitting results according to Equation (1) are displayed in Figure 8 as a dotted line. The fitting parameters and the calculated average decay time,

$$\tau_{dec} = \sum_{i=1}^{3} A_i \tau_i^2 / \sum_{i=1}^{3} A_i \tau_i, \quad (2)$$

are presented in Table 2. One can conclude that longer decay times are typical of perovskites on mesoporous TiO_2 (samples Film B and D lead against Film A and C, respectively) and of layers formed using the conventional spray pyrolysis method (samples Film A and B lead against Film C and D, respectively). Meanwhile, Film C demonstrates the shortest decay times that can be influenced by the carrier surface recombination, which correlates with the results of structure analysis. The triple exponential fitting model is not good enough in some cases. Thus, we should additionally consider the processes of trapping and de-trapping [14].

Photovoltage decay transients of the solar cells fabricated on different perovskite films are depicted in Figure 9. The photovoltage reaches its maximum value with a delay of approximately 100 to 200 ns after laser excitation. The dependence of photovoltage on time can be described by the following equation [13]:

$$V(t) = V_0 \left(\frac{e^{-t\tau_t} - e^{-t\tau_{rec}}}{1\tau_{rec} - 1\tau_t} \right) \quad (3)$$

where τ_t is the time constant of electron transport inside the perovskite layer, τ_{rec} is the electron recombination time, and V_0 is the initial photovoltage. Fitting results calculated according to Equation (3) are depicted in Figure 9 as a dotted line. The redistribution of excited charge carriers inside the perovskite layer is determined mainly by their diffusion since the electric field is weak here. Therefore, τ_t is approximately equal to L^2/D, where L is the layer thickness and D is the bipolar diffusion coefficient. The estimated D values are 0.12, 0.07, 0.25, and 0.07 cm^2/s for perovskite samples Film A, B, C, and D, respectively. These values of bipolar diffusion coefficient are typical of spin-coated perovskite layers [5,7,43–45]. It should be noted that the diffusion coefficient is higher in planar than in mesoporous structured perovskite solar cells, as was observed earlier in [44], whereas the lifetime of generated charge carriers is longer in mesoporous structured perovskite solar cells due to the better crystallinity of mesoporous structured perovskite films.

The current versus forward bias voltage characteristics of the solar cells fabricated on different perovskite films are presented in Figure 10. It can be seen that the mesoporous structured PSCs show better performance than the planar ones. The values of photovoltaic quantities of the best PSCs are presented in Table 3, and statistics of the power-conversion efficiency of 60 perovskite solar cells fabricated on different perovskite films are depicted in Figure 11. The high PCE of mesoporous structured perovskite solar cells is mainly determined by a high value of the short circuit current resulting from an enlarged perovskite/mp-TiO_2 interfacial area compared to the planar PSC case.

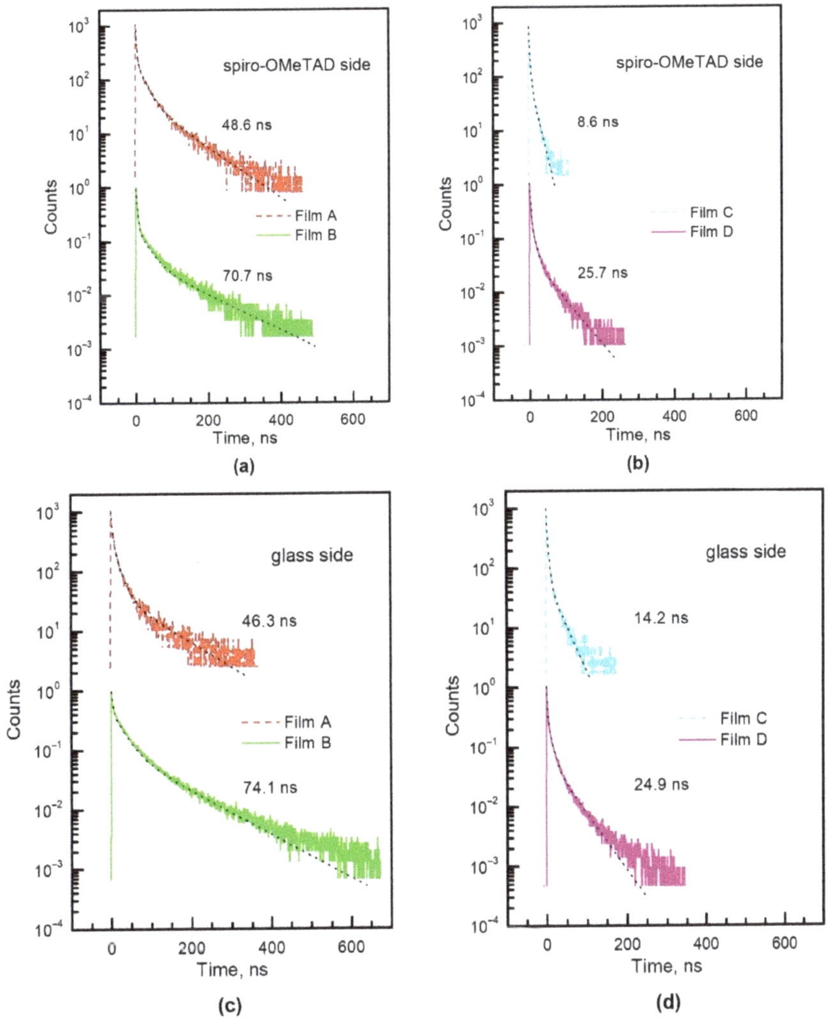

Figure 8. PL decay transients of PSC measured from the spiro-OMeTAD side (**a**,**b**) and from the glass side (**c**,**d**). Black dotted lines indicate triple exponent approximation fitting, and the numbers (in ns) are the averaged decay time constants.

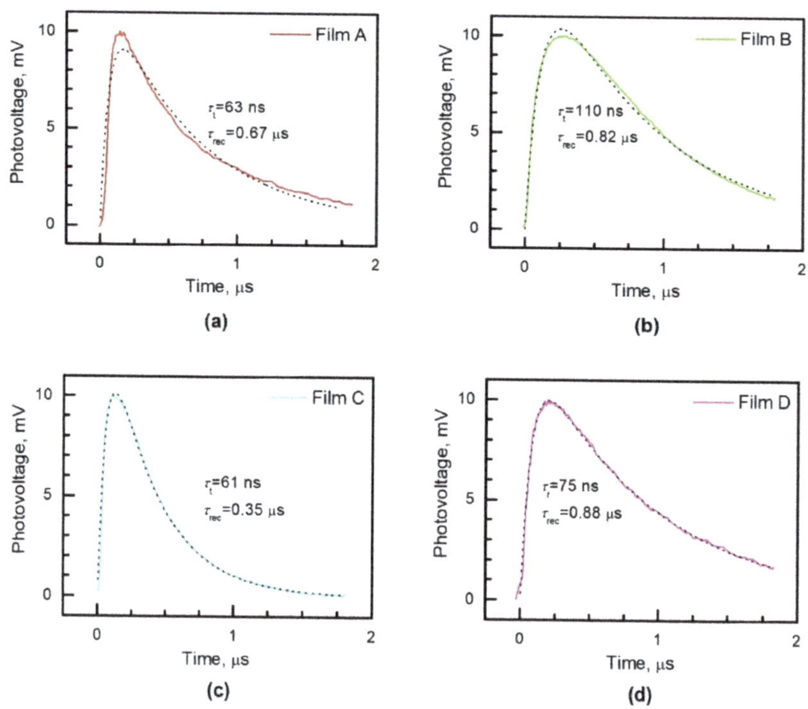

Figure 9. Photovoltage decay transients of the solar cells fabricated on the base of different perovskite films: (**a**) Film A, (**b**) Film B, (**c**) Film C, (**d**) Film D. Black dotted lines indicate fitting according to Equation (3), and τ_f and τ_{rec} are the electron transport and recombination time constants, respectively.

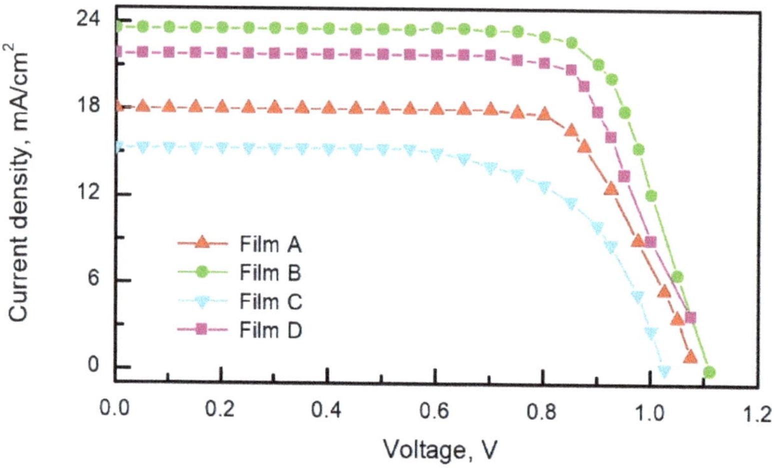

Figure 10. Current–voltage characteristics of the solar cells fabricated on different perovskite films under 100 mW/cm^2 spectral lamp irradiance.

Table 3. Photovoltaic parameters of the solar cells on the base of different perovskite films. V_{oc} is the open-circuit voltage; J_{sc} is the short circuit current; FF is the fill factor; PCE is the power-conversion efficiency.

Film	V_{oc}, V	J_{sc}, mA·cm^{-2}	FF, %	PCE, %
Film A	1.09	18	72	14.1
Film B	1.11	23.5	77	20.0
Film C	1.03	15.1	63	9.8
Film D	1.1	21.8	73	17.5

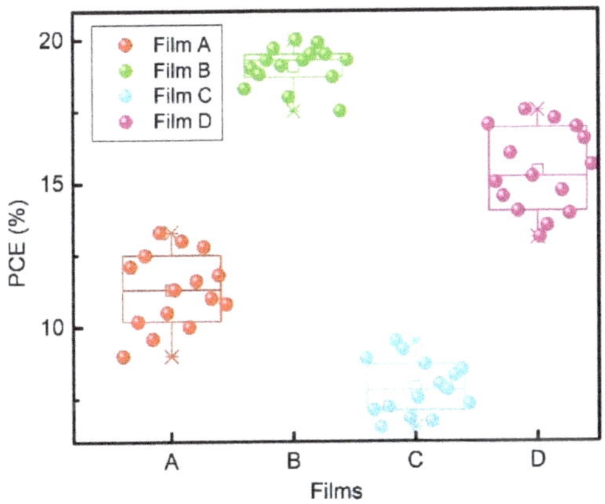

Figure 11. Statistical distribution of the power-conversion efficiency of perovskite solar cells based on different perovskite films.

It is worth noting that the high crystallinity of mesoporous structured perovskite also has a positive effect on the performance of PSCs because uniform perovskite films with high surface coverage and minimum defects are beneficial for the transportation of both electrons and holes [46]. Additionally, time-resolved photoluminescence decay and transient photovoltage measurements showed that the lifetime of generated charge carriers in mesoporous structured perovskite solar cells is longer than that in planar PSCs, and as a consequence, the performance of the mesoporous structured PSCs is better. Negligible hysteresis of the current–voltage characteristics of the perovskite $Cs_{0.1}(MA_{0.17}FA_{0.83})_{0.9}Pb(I_{0.83}Br_{0.17})_3$ solar cells is also a noteworthy fact, which was observed earlier in other works [33,37,47]. The planar PSCs fabricated on the base of Films A and C show the lowest PCE values, and this finding can be related to the shorter lifetime of charge carriers in these films resulting from their lower crystallinity (see Figures 8 and 9). On the other hand, the mesoporous structured perovskite solar cells fabricated on the base of Film B demonstrate the best PCE due to the high crystallinity of the perovskite film, giving a reason for the long lifetime of the generated charge carriers. In addition, the crystallites of Film B are the largest (see Table 1). This also greatly contributes to the improved performance of the mesoporous structured perovskite solar cells [47]. Thus, we see a good correlation between electronic processes in perovskite layers and the performance of perovskite solar cells.

4. Conclusions

An experimental study of the photoelectric properties of planar and mesoporous structured perovskite solar cells was conducted. It was established that the crystallinity of perovskite films grown on mesoporous titanium dioxide is better than that of films grown on a compact TiO_2. Transient photovoltage and time-resolved photoluminescence decay measurements show that the lifetime of charge carriers is also longer in mesoporous structured PSCs. The presented experimental data reveal the relationship between photovoltage decay characteristics and PSC performance, as both of them are affected by the mesoscopic structure of the layered structure. It was found that the mesoporous structured perovskite solar cells have a higher PCE than the planar ones, mainly due to an enlarged perovskite/mp-TiO_2 interfacial area and longer lifetime of generated charge carriers. These results provide a basis for improved further material growth and device fabrication to achieve better performance of perovskite solar cells.

Author Contributions: Conceptualization and methodology, S.A. and K.L.; sample fabrication, A.G., K.P., L.S. and A.Š.; formal analysis, A.S. (Algirdas Selskis), A.S. (Algirdas Sužiedėlis) and J.G.; experimental investigation, E.Š., K.P., A.Č., R.J. and A.L.; writing, reviewing and editing, S.A., J.G. and K.L.; visualization, A.S. (Algirdas Selskis) and K.P.; project administration and supervision, S.A. All authors have read and agreed to the published version of the manuscript.

Funding: This research was funded by the Research Council of Lithuania, grant number 01.2.2-LMT-K-718-01-0050.

Institutional Review Board Statement: Not applicable.

Informed Consent Statement: Not applicable.

Data Availability Statement: No new data were created or analyzed in this study. Data sharing is not applicable to this article.

Acknowledgments: We are sincerely thankful to Skaidra Bumelienė for her good will in the preparation of the article manuscript.

Conflicts of Interest: The authors declare no conflict of interest.

References

1. Mozaffari, N.; Walter, D.; White, T.P.; Bui, A.D.; Tabi, G.D.; Weber, K.; Catchpole, K.R. Unraveling the role of energy ban alignment and mobile ions on interfacial recombination in perovskite solar cells. *Sol. RRL* **2022**, *6*, 2101087. [CrossRef]
2. Kim, M.; Jeong, J.; Lu, H.; Lee, T.K.; Eickemeyer, F.T.; Liu, Y.; Choi, I.W.; Choi, S.J.; Jo, Y.; Kim, H.B.; et al. Conforma Quantum Dot-SnO_2 Layers as Electron Transporters for Efficient Perovskite Solar Cells. *Science* **2022**, *375*, 302–306. Availabl online: https://www.science.org/doi/10.1126/science.abh1885 (accessed on 24 April 2022). [CrossRef] [PubMed]
3. De Wolf, S.; Holovsky, J.; Moon, S.-J.; Löper, P.; Niesen, B.; Ledinsky, M.; Haug, F.-J.; Yum, J.-H.; Ballif, C. Organometallic halid perovskites: Sharp optical absorption edge and its relation to photovoltaic performance. *J. Phys. Chem. Lett.* **2014**, *5*, 1035–103 [CrossRef] [PubMed]
4. Quan, L.N.; Rand, B.P.; Friend, R.H.; Mhaisalkar, S.G.; Lee, T.W.; Sargent, E.H. Perovskites for next-generation optical source *Chem. Rev.* **2019**, *119*, 7444–7478. [CrossRef]
5. Wehrenfennig, C.; Eperon, G.E.; Johnston, M.B.; Snaith, H.J.; Herz, L.M. High carrier mobilities and lifetimes in organolea trihalide perovskites. *Adv. Mater.* **2014**, *26*, 1584–11589. [CrossRef]
6. Karakus, M.; Jensen, S.A.; D'Angelo, F.; Turchinovich, D.; Bonn, M.; Cánovas, E. Phonon-electron scattering limits free charg mobility in methylammonium lead iodide perovskites. *J. Phys. Chem. Lett.* **2015**, *6*, 4991–4996. [CrossRef]
7. Stranks, S.D.; Eperon, G.E.; Grancini, G.; Menelaou, C.; Alcocer, M.J.P.; Leijtens, T.; Herz, L.M.; Petrozza, A.; Snaith, H.J. Electro hole diffusion lengths exceeding 1 micrometer in an Organometal trihalide perovskite absorber. *Science* **2013**, *342*, 341–34 [CrossRef]
8. Dong, Q.; Fang, Y.; Shao, Y.; Mulligan, P.; Qiu, J.; Cao, L.; Huang, J. Electron-hole diffusion lengths > 175 μm in solution-grow $CH_3NH_3PbI_3$ single crystals. *Sci. Express* **2015**, *347*, 967–969. [CrossRef]
9. Kothandaraman, R.K.; Jiang, Y.; Feurer, T.; Tiwari, A.N.; Fu, F. Near-infrared-transparent perovskite solar cells and peroskite-base tandem photovoltaics. *Small Methods* **2020**, *4*, 2000395. [CrossRef]
10. Staub, F.; Hempel, H.; Hebig, J.C.; Mock, J.; Paetzold, U.W.; Rau, U.; Unold, T.; Kirchartz, T. Beyond bulk lifetimes: Insights int lead halide perovskite films from time-resolved photoluminescence. *Phys. Rev. Appl.* **2016**, *6*, 044017. [CrossRef]
11. Solanki, A.; Yadav, P.; Turren-Cruz, S.H.; Lim, S.S.; Saliba, M.; Sum, T.C. Cation influence on carrier dynamics in perovskite sola cells. *Nano Energy* **2019**, *58*, 604–611. [CrossRef]
12. Xie, Y.M.; Xue, Q.; Yip, H.L. Metal-halide perovskite krystallzation kinetics: A review of experimental and theoretical studie *Adv. Energy Mater.* **2021**, *11*, 2100784. [CrossRef]
13. Hydayat, R.; Nurunnizar, A.A.; Fariz, A.; Herman; Rosa, E.S.; Shobih; Oizumi, T.; Fujii, A.; Ozaki, M. Revealing the charg carrier kinetics in perovskite solar cells affected by mesoscopic stuctures and defect states from simple transient photovoltag measurements. *Sci. Rep.* **2020**, *10*, 19197. [CrossRef] [PubMed]
14. Pean, E.V.; Dimitrov, S.; de Castro, C.S.; Davies, M.L. Interpreting time-resolved photoluminescence of perovskite material *Phys. Chem. Chem. Phys.* **2020**, *22*, 28345. [CrossRef]

15. Lian, J.; Lu, B.; Niu, F.; Zeng, P.; Zhan, X. Electron-transport materials in perovskite solar cells. *Small Methods* **2018**, *2*, 1800082. [CrossRef]
16. Zhao, Y.; Ye, Q.; Chu, Z.; Gao, F.; Zhang, X.; You, J. Recent progress in high-efficiency planar-structure perovskite solar cells. *Energy Environ. Mater.* **2019**, *2*, 93–106. [CrossRef]
17. Liu, M.; Johnston, M.B.; Snaith, H.J. Efficient planar heterojunction perovskite solar cells by vapour deposition. *Nature* **2013**, *501*, 395–398. [CrossRef]
18. Kim, H.S.; Lee, C.R.; Im, J.H.; Lee, K.B.; Moehl, T.; Marchioro, A.; Moon, S.J.; Humphry-Baker, R.; Yum, J.H.; Moser, J.E.; et al. Lead iodide perovskite sensitized all-solid-state submicron thin film mesoscopic solar cell with efficiency exceeding 9%. *Sci. Rep.* **2012**, *2*, 591. [CrossRef]
19. Lee, M.M.; Teuscher, J.; Miyasaka, T.; Murakami, T.N.; Snaith, H.J. Efficient hybrid solar cells based on meso-superstructured organometal halide perovskites. *Science* **2012**, *338*, 643–647. [CrossRef]
20. Li, Y.; Cooper, J.K.; Buonsanti, R.; Giannini, C.; Liu, Y.; Toma, F.M.; Sharp, I.D. Fabrication of planar heterojunction perovskite solar cells by controlled low-pressure vapor annealing. *J. Phys. Chem. Lett.* **2015**, *6*, 493–499. [CrossRef]
21. Lee, D.G.; Kim, M.; Kim, B.J.; Kim, D.H.; Lee, S.M.; Choi, M.; Lee, S.; Jung, H.S. Effect of TiO_2 particle size and layer thickness on mesoscopic perovskite solar cells. *Appl. Surf. Sci.* **2019**, *477*, 131–136. [CrossRef]
22. Kim, J.Y.; Lee, J.W.; Jung, H.S.; Shin, H.; Park, N.G. High-efficiency perovskite solar cells. *Chem. Rev.* **2020**, *120*, 7867–7918. [CrossRef] [PubMed]
23. Wang, J.; Zou, X.; Zhu, J.; Cheng, J.; Chen, D.; Bai, X.; Yao, Y.; Chang, C.; Yu, X.; Liu, B.; et al. Effect of optimization of TiO_2 electron transport layer on performance of perovskite solar cells with rough FTO substrates. *Materials* **2020**, *13*, 2272. [CrossRef] [PubMed]
24. Marchioro, A.; Teuscher, J.; Friedrich, D.; Kunst, M.; Krol, R.; Moehl, T.; Grätzel, M.; Moser, J.-E. Unravelling the mechanism of photoinduced charge transfer processes in lead iodide perovskite solar cells. *Nat. Photonics* **2014**, *8*, 250–255. [CrossRef]
25. Starowicz, Z.; Gawlińska, K.; Walter, J.; Socha, R.P.; Kulesza-Matlak, G.; Lipiński, M. Extended investigation of sol aging effect on TiO_2 electron transporting layer and performances of perovskite solar cells. *Mater. Res. Bull.* **2018**, *99*, 136–143. [CrossRef]
26. Sławek, A.; Starowicz, Z.; Lipiński, M. The influence of the thickness compact TiO_2 electron transporting layer on the performances of planar $CH_3NH_3PbI_3$ perovskite solar cells. *Materials* **2021**, *14*, 3295. [CrossRef]
27. Xiao, M.; Huang, F.; Huang, W.; Dkhissi, Y.; Zhu, Y.; Etheridge, J.; Gray-Weale, A.; Bach, U.; Cheng, Y.-B.; Spiccia, L. A fast deposition-crystallization procedure for highly efficient lead iodide perovskite thin-film solar cells. *Angew. Chem. Int. Ed. Engl.* **2014**, *53*, 9898–9903. [CrossRef]
28. Saliba, M.; Correa-Baena, J.-P.; Wolff, C.M.; Stolterfoht, M.; Phung, N.; Albrecht, S.; Neher, D.; Abate, A. How to make over 20% efficient perovskite solar cells in regular (n–i–p) and inverted (p–i–n) architectures. *Chem. Mater.* **2018**, *30*, 4193–4201. [CrossRef]
29. Grätzel, M. The light and shade of perovskite solar cells. *Nat. Mater.* **2014**, *13*, 838–842. Available online: www.nature.com/articles/nmat4065 (accessed on 25 April 2022). [CrossRef]
30. Yang, Y.; Ri, K.; Mei, A.; Liu, L.; Hu, M.; Liu, T.; Li, X.; Han, H. The size effect of TiO_2 nanoparticles on a printable mesoscopic perovskite solar cell. *J. Mater. Chem.* **2015**, *3*, 9103–9107. [CrossRef]
31. Hu, H.; Dong, B.; Hu, H.; Chen, F.; Kong, M.; Zhang, Q.; Luo, T.; Zhao, L.; Guo, Z.; Li, J.; et al. Atomic layer deposition of TiO_2 for a high-efficiency hole-blocking layer in hole-conductor-free perovskite solar cells processed in amber air. *Appl. Mater. Interfaces* **2016**, *8*, 17999–18007. [CrossRef] [PubMed]
32. Sun, X.; Xu, J.; Xiao, L.; Chen, J.; Zhang, B.; Yao, J.; Dai, S. Influence of the porosity of the TiO_2 film on the performance of the perovskite solar cells. *Int. J. Photoenergy* **2017**, *2017*, 4935265. [CrossRef]
33. Ašmontas, S.; Čerškus, A.; Gradauskas, J.; Grigucevičienė, A.; Leinartas, K.; Lučun, A.; Petrauskas, K.; Selskis, A.; Sužiedelis, A.; Širmulis, E.; et al. Cesium-containing triple cation perovskite solar cells. *Coatings* **2021**, *11*, 279. [CrossRef]
34. Ašmontas, S.; Čerškus, A.; Gradauskas, J.; Grigucevičienė, A.; Juškėnas, R.; Leinartas, K.; Lučun, A.; Petrauskas, K.; Selskis, A.; Sužiedelis, A.; et al. Impact of cesium concentration on optoelectronic properties of metal halide perovskites. *Materials* **2022**, *15*, 1936. [CrossRef]
35. Ašmontas, S.; Anbinderis, M.; Gradauskas, J.; Juškėnas, R.; Leinartas, K.; Lučun, A.; Selskis, A.; Staišiūnas, L.; Stanionytė, S.; Sužiedėlis, A.; et al. Low resistance TiO_2/p-Si heterojunction for tandem solar cell. *Materials* **2020**, *13*, 2857. [CrossRef]
36. Kundrotas, J.; Čerškus, A.; Valušis, G.; Linfield, E.H.; Johannessen, E.; Johannessen, A. Dynamics of free carriers-neutral impurity related optical transitions in Be and Si δ-doped GaAs/AlAs multiple quantum wells: Fractional-dimensional space approach. *Lith. J. Phys.* **2014**, *54*, 233–243. [CrossRef]
37. Saliba, M.; Matsui, T.; Seo, J.Y.; Domaski, K.; Correa-Baena, J.-P.; Nazeeruddin, M.K.; Zakeeruddin, S.M.; Tress, W.; Abate, A.; Hagfeldt, A.; et al. Cesium-containing triple cation perovskite solar cells: Improved stability, reproducibility and high efficiency. *Energy Environ. Sci.* **2016**, *9*, 1989–1997. [CrossRef]
38. Schmidt, T.; Lischka, K.; Zulehner, W. Excitation-power dependence of the near-band-edge photoluminescence of semiconductors. *Phys. Rev. B* **1992**, *45*, 8989–8994. [CrossRef]
39. Shibata, H.; Sakai, M.; Yamada, A.; Matsubara, K.; Sakurai, K.; Tampo, H.; Ishizuka, S.; Kim, K.-K.; Niki, S. Excitation-power dependence of free exciton photoluminescence of semiconductors. *Jpn. J. Appl. Phys.* **2005**, *44*, 6113–6114. [CrossRef]
40. Spindler, C.; Galvani, T.; Wirtz, L.; Rey, G.; Siebentritt, S. Excitation-intensity dependence of shallow and deep-level photoluminescence transitions in semiconductors. *J. Appl. Phys.* **2019**, *126*, 175703–175709. [CrossRef]

41. Dursun, I.; Zheng, Y.; Guo, T.; de Bastiani, M.; Turedi, B.; Sinatra, L.; Haque, M.A.; Sun, B.; Zhumekenov, A.A.; Saidaminov, M.I.; et al. Efficient photon recycling and radiation trapping in cesium lead halide perovskite waveguides. *ACS Energy Lett.* **2018**, *3*, 1492–1498. [CrossRef]
42. Ca, N.X.; Hien, N.T.; Luyen, N.T.; Lien, V.T.K.; Thanh, L.D.; Do, P.V.; Bau, N.Q.; Pham, T.T. Photoluminescence properties of CdTe/CdTeSe/CdSe core/alloyed/shell type-II quantum dots. *J. Alloys Compd.* **2019**, *787*, 823–830. [CrossRef]
43. Leijtens, T.; Stranks, S.D.; Eperon, G.E.; Lindblad, R.; Johansson, E.M.J.; McPherson, I.J.; Rensmo, H.; Ball, J.M.; Lee, M.M.; Snaith, H.J. Electronic properties of meso-superstructured and planar organometal halide perovskite films:charge trapping, photodoping, and carrier mobility. *ACS Nano* **2014**, *8*, 7147–7155. [CrossRef] [PubMed]
44. Hutter, E.M.; Eperon, G.E.; Stranks, S.D.; Savenije, T.J. Charge carrier in planar and meso-structured organic-inorganic perovskites: Mobilities, lifetimes, and concentrations of trap states. *J. Phys. Chem. Lett.* **2015**, *6*, 3082–3090. [CrossRef] [PubMed]
45. Ščajev, P.; Qin, C.; Aleksiejūnas, R.; Baronas, P.; Miasojedovas, S.; Fujihara, T.; Matsushima, T.; Adachi, C.; Juršėnas, S. Diffusion enhancement in highly excited $MAPbI_3$ perovskite layers with additives. *J. Phys. Chem. Lett.* **2018**, *9*, 3167–3172. [CrossRef]
46. Li, Y.; Ji, L.; Liu, R.; Zhang, C.; Mak, C.H.; Zou, X.; Shen, H.H.; Leu, S.Y.; Hsu, H.Y. A revew on morphology engineering for highly efficient and stable hybrid perovskite solar cells. *J. Mater. Chem. A* **2018**, *6*, 12842–12875. [CrossRef]
47. Singh, T.; Miyasaka, T. Stabilizing the efficiency beyond 20% with a mixed cation perovskite solar cell in ambient air under controlled humidity. *Adv. Energy Mat.* **2018**, *8*, 1700677. [CrossRef]

materials

MDPI

Article

Numerical Study on Hydrodynamic Characteristics and Electrochemical Performance of Alkaline Water Electrolyzer by Micro-Nano Surface Electrode

Ye Xia [1], Mengyu Gao [1,*], Jincheng Yu [1], Yang Si [1,2,*], Laijun Chen [1,2] and Shengwei Mei [1,2]

1 New Energy (Photovoltaic) Industry Research Center, Qinghai University, Xining 810016, China; 18081291358@163.com (Y.X.); yjc18636285020@163.com (J.Y.); chenlaijun@tsinghua.edu.cn (L.C.); meishengwei@tsinghua.edu.cn (S.M.)
2 State Key Laboratory of Control and Simulation of Power System and Power Generation Equipment, Electric Machinery Department, Tsinghua University, Beijing 100084, China
* Correspondence: mygao16@fudan.edu.cn (M.G.); siyang@qhu.edu.cn (Y.S.)

Abstract: This study constructed a two-dimensional alkaline water electrolyzer model based on the two-phase flow Euler–Euler model. In the model, the micro-nano surface electrodes with different structure types and graphic parameters (distance, height, and width) were used and compared with the vertical flat electrode to evaluate their influence on electrolysis performance. The simulation results show that the performance of the micro-nano surface electrode is much better than that of the vertical flat electrode. The total length of micro-nano structural units relates to the contact area between the electrode and the electrolyte and affects the cell voltage, overpotential, and void fraction. When rectangular structural units with a distance, height, and width of 0.5 μm, 0.5 μm, and 1 μm are used, the total length of the corresponding micro-nano surface electrode is three times that of the vertical flat electrode, and the cathode overpotential decreases by 65.31% and the void fraction increases by 54.53% when it replaces the vertical flat electrode.

Keywords: alkaline water electrolysis; two-phase flow; micro-nano fabrication; micro-nano surface electrode; numerical simulation

1. Introduction

Increased demand for energy, depletion of fossil fuels, and global warming have become global challenges recently; thus, there is an urgent need for renewable energy development [1]. Among all the alternatives to fossil fuels, hydrogen is one of the most promising clean energies due to its abundant reserves and high energy density [2]. Hydrogen renewable energy system includes hydrogen production, storage, and utilization [3]. In the process of hydrogen production, high-purity hydrogen produced by water electrolysis is favored by hydrogen-based fuel cells [4]. Water electrolysis is divided into alkaline water electrolysis, proton exchange membrane, and solid oxide electrolysis [5,6]. Hydrogen production from alkaline water electrolysis is closest to industrial application, and a traditional alkaline water electrolyzer (AWE) mainly includes a cathode, anode, diaphragm, and electrolyte. Among them, the activation energy barrier and catalytic active area of the electrode determine the electrolysis efficiency [7–10].

Experimentally, as early as 1982, researchers realized that the surface roughness of an alloy catalyst is higher than that of pure metal and has more catalytic active sites. Brown et al. [11] prepared the electrodes with a low overvoltage for hydrogen evolution and good stability. The electrodes with a larger catalytic surface area can reduce the overvoltage in the 70–90 mV range at 70 °C and the current density of 1 A cm^{-2}. In 2020, Zhang et al. [12] synthesized Ni_3S_2 three-dimensional rod array structures doped with trace amounts of Fe. Experiments showed that the catalyst has excellent catalytic activity and corrosion resistance in acidic, neutral, and alkaline solutions. In the 1.0 M KOH solution, only small overpotentials of 105 mV and 213 mV were required to achieve the current density of 10 mA cm^{-2} for hydrogen evolution reaction (HER) and oxygen evolution reaction (OER), respectively. In 2021, Wang et al. [13] synthesized an IrNi-$FeNi_3$ hybrid self-supported on nickel foam with unique nanosheets, and the results showed that only 1.47 V is required for overall water splitting to arrive at 10 mA cm^{-2}. They found that the unique self-supported nanosheets structure on nickel foam can expose more active sites, and the hybridization of IrNi with $FeNi_3$ can promote the intrinsic activity of catalyst. According to the above, it can note that increasing the catalytic surface area can reduce the overpotential of hydrogen production in actual experiments.

In recent years, many numerical studies have been carried out on AWE modeling, characterization, and analysis, mainly accomplished by coupling the current distribution model and computational fluid dynamics (CFD) model [14]. Rodríguez et al. [15] established a two-dimensional (2D) CFD model of AWE, which considered and analyzed the electrochemical phenomena and fluid dynamics involved in AWE under different operating conditions and verified them by experiments. For all cases, the model fitted very well with the experimental data, and the error of the polarization curve was less than 1%. Zarghami et al. [16] used the Euler–Euler model to simulate multiphase flow in AWE. Their simulation results compared with the experimental data and were in reasonable agreement with the experimental data at a higher current density due to the turbulence dispersion force added. Liu et al. [17] constructed a two-phase bubble flow 2D numerical model in the electrolytic cell to study hydrodynamics and mass transfer based on the Euler–Euler method. The results showed that the effects of different operating conditions on the volume fraction and bubble layer were in good agreement with the experimental results. According to the above, it can note that with the development of computer science, the numerical simulation of hydrogen production by water electrolysis has gradually become a research hotspot [18–20]. Compared with the actual experiment, one of the advantages of numerical simulation is that it can get relevant results without actual experiments, which is cost-effective and has a low time cost for experimental design.

As mentioned above, increasing the catalytic surface area is an important method to improve hydrogen production efficiency by water electrolysis [21,22]. Meanwhile, surface patterning is essential to improve the surface area of materials [23–25]. Therefore, it is a crucial way to design the electrode surface to obtain the influence rule of pattern parameters on the efficiency of hydrogen production by water electrolysis through numerical simulation, especially for "top-down" construction approaches such as lithography [26,27], which can reduce its design cost. However, there are few simulation studies on the influence of micro-nano structure graphic parameters on hydrogen production by water electrolysis. Therefore, this study numerically investigates the impact of micro-nano structures on the electrochemical and fluid performance of AWE based on the Euler–Euler method. The influence factors of the micro-nano structure were added under the steady-state, 2D, and two-phase flow CFD model considering turbulence.

2. Model

2.1. Physical Model

The 2D geometric model of the AWE referring to commercial AWE is shown in Figure 1 [28,29]. The AWE is divided into cathodic and anodic compartments by a separator known as a diaphragm, which is to avoid the mixing of the two gases and maintains a low resistivity. These compartments are submerged in the electrolyte, an aqueous potassium hydroxide (KOH) solution. The cathode and the anode are located on the left and right walls of the gas compartment, respectively. The hydrogen gas (H_2) evolves at the cathode, whereas oxygen gas (O_2) is generated at the anode. The electrochemical reactions are shown in Equation (1) and Equation (2), respectively.

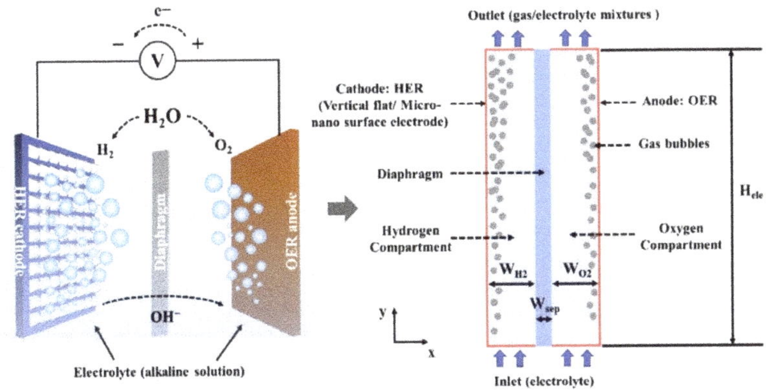

Figure 1. Schematic diagram of the 2D geometric model of AWE.

For the cathode, a vertical flat electrode was changed into a micro-nano surface electrode to investigate the influence of micro-nano structures on the AWE performance in this study. The design flow of micro-nano surface electrodes is shown in Figure 2. Firstly, six structures are mainly selected, namely rectangular (r), triangle (p), trapezoidal (t), inverted rectangular (ir), inverted triangle (ip), and inverted trapezoidal (it). Then the graphic parameters (distance, height, width) are selected to construct the structural units. Thirdly, the micro-nano surface electrodes are formed. Finally, the micro-nano surface electrodes are applied to the AWE model and simulated.

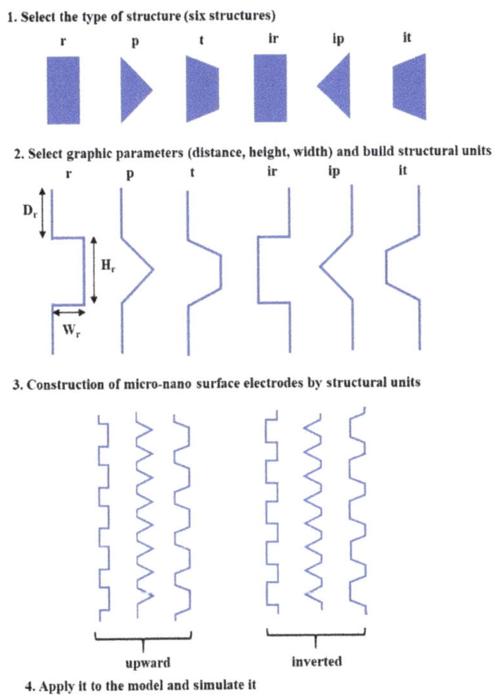

Figure 2. Design flow of micro-nano surface electrodes.

The electrode surfaces are set at the height of 10 mm, the same as the diaphragm. The diaphragm is a thin rectangle with a width of 1 mm. The cathodic and anodic compartments are described as thick rectangles with a width of 2 mm. The KOH electrolyte enters from the inlet, and H_2 and O_2 bubbles evolve on the cathode and anode, respectively. The generated bubbles move along the x and y axes to form a bubble layer, and the resulting two-phase mixture (gas and electrolyte) leaves from the outlet.

$$2H_2O(l) + 2e^- = H_2(g) + 2OH^- \tag{1}$$

$$4OH^- = 2H_2O(l) + O_2(g) + 4e^- \tag{2}$$

2.2. Numerical Model

Multiphase flow models are broadly classified into Euler–Euler and Euler–Lagrange approaches. The Euler–Lagrange model is challenging to deal with in the large number of bubbles that consume a long computing time. Moreover, the Euler–Lagrange model generally does not consider the bubble–bubble interactions and ignores the "numerical volume" occupied by bubbles [30,31]. Hence, it is not suitable for this study to consider turbulent dispersion force. The Euler–Euler model can handle numerous types of multiphase flow and is suitable for simulating the macroscopic behavior of many bubbles rising through the liquid [32]. Consequently, the Euler–Euler model is generally considered the most suitable for the two-phase flow.

As previously mentioned, the electrochemical phenomena and fluid characteristics of AWE are numerically simulated using COMSOL Multiphysics 5.6. In this study, the model was built using the "water electrolyzer" module to calculate the current distribution and plot the polarization curve. The "Euler–Euler model, turbulent flow" module was used to study the generation and distribution of two gases in compartments. This model was solved in a steady-state.

The model has been designed using the following assumptions to ensure a manageable problem [33–35]:

1. The fluids in both phases are Newtonian, viscous, and incompressible;
2. The physical properties remain constant;
3. The two phases share the same pressure field;
4. The electrolyte is considered distributed uniformly due to the distribution of the ions has little effect on the result;
5. The flow is considered isothermal. Thus, heat exchange and energy equation are not considered;
6. The influence of surface tension is negligible;
7. No bubble coalescence or break-up occurs, and the bubble diameter can be considered constant at a given current density;
8. The bubble–bubble interaction is considered by introducing the turbulent dispersion force;
9. Although the Reynolds number is small, the fluid is considered laminar flow. However, considering the observed turbulence phenomenon by Boissonneau et al. [36] and Aldas et al. [37], a simple turbulence model is selected to describe the turbulence behavior in the study;
10. Since alkaline electrolysis is a complex physical problem involving two-phase flow and the construction of micro-nano surface electrodes is complicated, all issues are simplified to 2D and steady-state simulation.

2.2.1. Secondary Current Distribution Model

Secondary current distribution is used to simulate the electrical property of AWE in this work. Equation (3) shows the total potential obtained by simulation [38]. Based on the above assumptions, Ohm's law, current conservation, and the Butler–Volmer equation are used to solve the current density distribution and explain the influence of electrode kinetics, as shown in Equation (4), Equation (5), and Equation (6), respectively [39–42].

$$U_{cell} = E_{rev} + \eta_{ohm} + \eta_{act,a}(j) + \eta_{act,c}(j) \tag{3}$$

Ohm's law:

$$\vec{j} = -\sigma \nabla \varphi \tag{4}$$

Current conservation:

$$\nabla \cdot \vec{j} = Q \tag{5}$$

Butler–Volmer equation:

$$j_{loc} = j_0 \left(\exp\left(\frac{\alpha_a F \eta}{RT}\right) - \exp\left(-\frac{\alpha_c F \eta}{RT}\right) \right) \tag{6}$$

where E_{rev} is the reversible voltage in V, η_{ohm} is the ohmic overpotential in V, $\eta_{act,a}$ and $\eta_{act,c}$ is the activation overpotential at anode and cathode respectively in V, $\vec{j}$ is the current density vector in A m^{-2}, φ is the electrical potential in V, σ is the conductivity in S m^{-1}, Q is a general current source term in A m^{-3}, j_{loc} is the local charge transfer current density in A m^{-2}, j_0 is the exchange current density in A m^{-2}, α_a and α_c are the anodic and cathodic charge transfer coefficient, respectively, η is the activation overpotential in V, F is the Faraday constant in C mol^{-1}, and R is the universal gas constant in J mol^{-1} K^{-1}.

2.2.2. CFD Model

The Euler–Euler model is applied to represent the two-phase flow behavior. The model calculates the void fraction of each phase in two phases without defining each bubble in detail. Each phase has a velocity field, and the dynamics of each phase are described by the momentum equation and the continuity equation, as shown in Equation (7) and Equation (8), respectively. Moreover, the volume fractions are assumed continuous functions of space and time, and their sum is equal to one, i.e., $\alpha_g + \alpha_l = 1$ [34].

Momentum equation:

$$\frac{\partial}{\partial t}(\rho_k \alpha_k \vec{u}_k) + \nabla(\rho_k \alpha_k \vec{u}_k \vec{u}_k) = \alpha_k \rho_k \vec{g} - \alpha_k \nabla p + \nabla(\alpha_k \tau_k) + \vec{F}_k \tag{7}$$

Continuity equation:

$$\frac{\partial}{\partial t}(\alpha_g \rho_g + \alpha_l \rho_l) + \nabla(\alpha_g \rho_g \vec{u}_g + \alpha_l \rho_l \vec{u}_l) = 0 \tag{8}$$

where u is the velocity in m s^{-1}, p is the pressure in Pa, α is the void fraction, ρ is the density in kg m^{-3}, g is the gravitational acceleration in m s^{-2}, τ is the stress tensor in N m^{-2}, F_k is the volume force in N m^{-3}, and the subscript k refers to gas phase (g) or liquid phase (l).

The transport equation of the void fraction of gas is shown in Equation (9).

$$\frac{\partial \alpha_g \rho_g}{\partial t} + \nabla(\alpha_g \rho_g \vec{u}_g) = -\dot{m}_{gl} \tag{9}$$

where $\dot{m}_{gl}$ is the mass transfer rate from gas to liquid in g m^{-3} s^{-1}.

In this simplified 2D model, the gas flow generated (g m^{-2} s^{-1}) on the active surface of the electrode is defined by the Faraday equation, as shown in Equation (10).

$$\dot{m}_g = \frac{j \cdot M_g}{nF} \tag{10}$$

where j is the current density in A m^{-2}, M_g is the molar mass of the gas in g mol^{-1}, n is the number of transferred electrons, and the subscript g refers to H_2 or O_2.

Although the Reynolds number range is low, referring to the turbulence phenomenon observed by Mat et al. [37,43], and to better characterize the fluid flow behavior, therefore, a simple turbulence model was used. For turbulence, the k-ε model is applied. The stress tensor τ and the turbulent viscosity μ_T (Pa s) are shown in Equation (11) and Equation (12), respectively. The k-ε model solves two additional transport equations for two additional variables: the turbulent kinetic energy k (m^2 s^{-2}) and the turbulent dissipation rate ε (m s^{-3}), as shown in Equation (13) and Equation (14), respectively [44].

$$\tau_k = (\mu_k + \mu_T)(\nabla\vec{u}_k + (\nabla\vec{u}_k)^T - \frac{2}{3}(\nabla\cdot\vec{u}_k)I) - \frac{2}{3}\rho_k kI \tag{11}$$

$$\mu_T = \rho_k C_\mu \frac{k}{\varepsilon} \tag{12}$$

$$\rho\frac{\partial k}{\partial t} + \rho\vec{u}\nabla k = \nabla\cdot((\mu + \frac{\mu_T}{\sigma_k})\nabla k) + P_k - \rho\varepsilon \tag{13}$$

$$\rho\frac{\partial \varepsilon}{\partial t} + \rho\vec{u}\nabla\varepsilon = \nabla((\mu + \frac{\mu_T}{\sigma_k})\nabla\varepsilon) + C_{\varepsilon,1}\frac{\varepsilon}{k}P_k - C_{\varepsilon,2}\rho\frac{\varepsilon^2}{k} \tag{14}$$

where μ is the viscosity in Pa s, I is the unit tensor in N m^{-2}, the subscript k refers to gas phase (g) or liquid phase (l), C_μ, $C_{\varepsilon,1}$, $C_{\varepsilon,2}$, and σ_k are all constants in the turbulence model, and P_k is related to bubble-induced turbulence.

In addition to the drag force hindering the flow in the AWE, in fact, with the continuous generation and accumulation of bubbles, the collision between bubbles increases, which triggers the diffusion of bubbles from a high concentration to a low concentration. Therefore, based on previous work, drag force and turbulent dispersion force are considered in the model, as shown in Equation (15) and Equation (16), respectively [45,46].

$$\vec{F}_D = -\frac{3}{4}\rho_k\alpha_k\frac{C_d}{d_b}\left|\vec{U}_r\right|\vec{U}_r \tag{15}$$

$$\vec{F}_{BD} = -\alpha_k\rho_k\frac{K_g}{d_b}\left|\vec{U}_r\right|\nabla\alpha_k \tag{16}$$

where C_d is the drag coefficient, K_g is the diffusion factor of mixed bubbles in m s^{-2}, and d_b is the bubble diameter in m. The drag coefficient is defined by the Schiller–Naumann model as shown in Equation (17).

$$C_d = \begin{cases} \frac{24}{Re}(1 + 0.15Re^{0.687}) & Re \leq 1000 \\ 0.44 & Re > 1000 \end{cases} \tag{17}$$

where Re is the Reynolds number of bubbles calculated according to relative velocity, $Re = \rho_l U \rightarrow_r d_b/\mu_l$.

2.2.3. Initial Values and Boundary Conditions

The boundary conditions of the electrochemical and hydrodynamic models are shown in Figure 3. For the electrochemical model, the top and bottom boundaries of the model are electric insulation. For the hydrodynamic model, the velocity inlet and pressure outlet are used as the boundary conditions at the inlet and the outlet. No slip is used as the boundary condition of electrodes and the boundaries of the diaphragm. Their velocity is set to 0 m s^{-1}. At the electrode surfaces, gas and liquid mass flux boundary conditions are used. At the boundaries of the diaphragm, the liquid phase mass flux corresponding to the ionic flux of OH^- from the current distribution model is adopted.

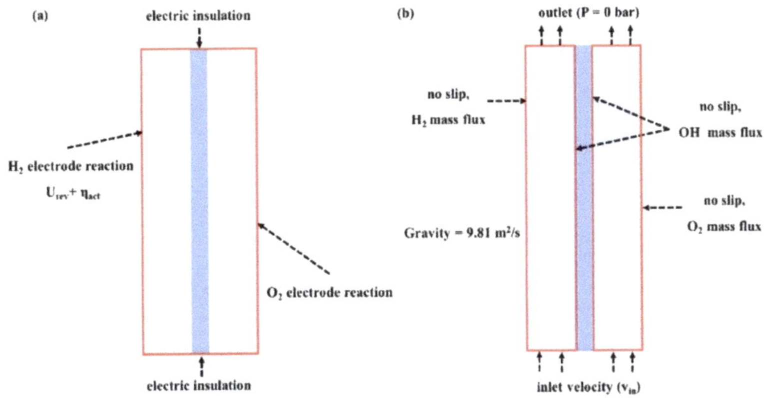

Figure 3. Boundary conditions of AWE. (**a**) Electrochemical model; (**b**) Hydrodynamic model.

Because of the bubble effect, the conductivity of the KOH electrolyte will decrease, and the generated bubbles will cover the electrode surface, affecting the exchange current density of the electrode reaction. Therefore, the effective electrolyte conductivity and the effective exchange current density are described according to the Bruggeman equation, as shown in Equation (18) and Equation (19), respectively [28,33].

$$\sigma_{eff} = \sigma_0 (1 - \alpha_g)^{1.5} \tag{18}$$

$$i_{eff} = i_0 (1 - \alpha_g)^{1.5} \tag{19}$$

where σ_{eff} is the effective electrolyte conductivity in S m^{-1}, σ_0 is the total electrolyte conductivity in S m^{-1}, i_{eff} is the effective exchange current density in A m^{-2}, i_0 is the exchange current density in A m^{-2}, α_g is the gas volume fraction, and the subscript g refers to H_2 or O_2.

The initial values for solving the model are shown in Table 1.

Table 1. Initial values of the 2D AWE model.

Symbol	Value	Unit	Description
R	8.314	J mol^{-1} K^{-1}	Universal gas constant
F	96,485.3	C mol^{-1}	Faraday constant
W_{H2}	2	mm	Hydrogen compartment width
W_{sep}	1	mm	Diaphragm width
W_{O2}	2	mm	Oxygen compartment width
H_{elec}	1	cm	Electrode height
W_r	0.5, 1	µm	Structure width
H_r	0.5, 1	µm	Structure height
D_r	0.5, 1	µm	Structure distance
d_{bubble}	50, 200	µm	Bubble diameter
T	70	°C	Operating temperature
p	1	atm	Pressure
i_{0H2}	100	A m^{-2}	Exchange current density, hydrogen oxidation
i_{0O2}	1	A m^{-2}	Exchange current density, oxygen reduction
M_{H2O}	18	g mol^{-1}	Water molar mass
M_{H2}	2	g mol^{-1}	Hydrogen molar mass
M_{O2}	32	g mol^{-1}	Oxygen molar mass
M_{OH}	17	g mol^{-1}	Hydroxide molar mass
c_{KOH}	3	M	Electrolyte concentration
v_{in}	0.2	m s^{-1}	Average inlet velocity

2.3. Grid Study and Validation

In this study, a physics-controlled mesh was used to adopt the modules, and a triangular mesh was generated. Mesh elements were fine predefined on the electrolyte domain and extra fine on the electrodes and diaphragm boundaries to correctly handle the effects in these contours. For details of grid and model study and validation, please refer to Table S1, Figures S1–S3 of Supplementary Materials.

3. Results

3.1. Vertical Flat Electrode

The bubble size produced in water electrolysis is affected by multiple factors, such as current density, electrode, and electrolyte flow rate. Therefore, two constant bubble diameters are considered in this paper to fit the actual situation as far as possible. According to the literature results [47,48], the bubble diameters produced by the vertical flat electrode and micro-nano surface electrode are 200 μm and 50 μm, respectively. The effects of these two bubble diameters were simulated using the vertical flat electrode to show the influence of bubble diameter change on electrolysis performance.

The gas void fraction distribution corresponding to two bubble diameters at different current densities is shown in Figure 4. It is seen that the bubble layer is parabolic distribution, and the void fraction increases with the increase of current density because the electrochemical reaction rate is proportional to the current density. With the increase of current density, the gas production rate increases, which leads to the increase of void fraction, thus promoting the thickening of the bubble layer. Moreover, the high void fraction will lead to more turbulent dispersion, which will distance the bubbles far from the electrode, expand the bubble layer, and increase the local void fraction by increasing the total void fraction.

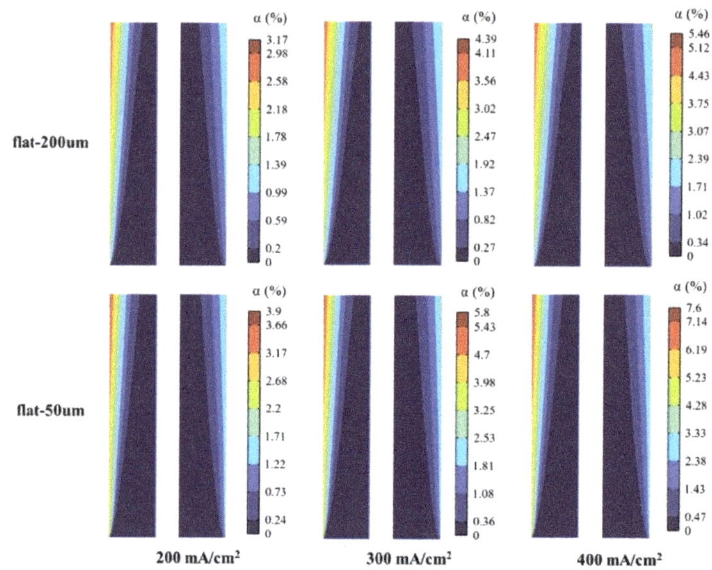

Figure 4. Gas void fraction distribution corresponds to two bubble diameters under different current densities.

Gas void fraction curves of these two bubble diameters at the top of the H_2 compartment under different current densities are shown in Figure 5. Gas void fraction curves of these two bubble diameters on vertical flat electrodes under different current densities are shown in Figure 6. Bubbles diffuse along the x-direction, which is due to the different gas concentrations in the electrolyte. Meanwhile, the bubbles aggregate at the bottom of the cathode, increasing the void fraction and promoting the bubble diffusion along the y-direction. From a certain height of the electrode, the void fraction decreases along the y-direction, which is caused by the movement of bubbles away from the electrode.

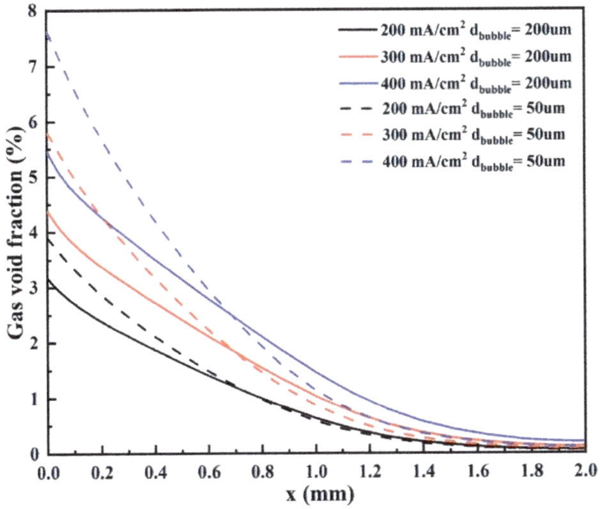

Figure 5. Gas void fraction curves of these two bubble diameters at the top of the H_2 compartment under different current densities.

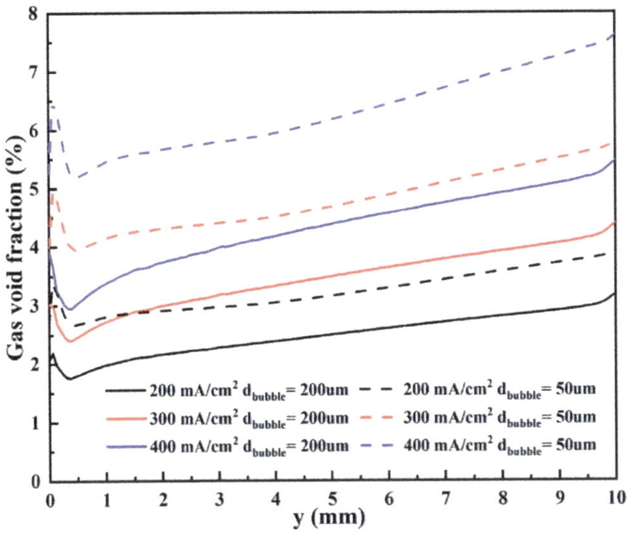

Figure 6. Gas void fraction curves of these two bubble diameters on vertical flat electrodes under different current densities.

For these two bubble sizes, the motion behavior of bubbles and the change of void fraction with current density are consistent. However, a smaller bubble size will lead to a higher void fraction. As shown in these figures, when the bubble diameter is 50 μm (flat-50), the void fraction on the electrode surface is higher than that of 200 μm (flat-200) at any current density. Moreover, as described in the literature [30], the thickness of the bubble layer will increase through larger bubbles, which is why the void fraction of bubble diameter of 200 μm will be slightly more significant after about 0.7 mm in Figure 5. Therefore, the trend that generating such tiny bubbles is beneficial to improving the efficiency of hydrogen production by electrolysis of water is consistent with the literature [47,48].

3.2. Micro-Nano Surface Electrode

3.2.1. Polarization Curve

The overpotentials of micro-nano surface electrodes with different structure types and graphic parameters under the same current density of 10 mA cm^{-2} are shown in Figure 7. It is seen that the overpotentials of p and ip, t and it, r, and ir are similar under the same graphic parameters, and the overpotentials of r and ir are the lowest, followed by t and it. Moreover, by sorting the overpotential from small to large, it is found that the corresponding graphic parameters of each structure are the same under the same ranking. For example, for all the structures, the overpotential is the smallest when the distance, height, and width are 0.5 μm, 0.5 μm, and 1 μm, respectively. However, the overpotentials of micro-nano surface electrodes with the same structure type and different graphic parameters are the same. For example, taking a rectangle as an example, their overpotentials are the same at graphic parameters of 1 μm, 1 μm, and 1 μm and 0.5 μm, 0.5 μm, and 0.5 μm. The differences between these micro-nano surface electrodes are reflected in the upward-inverted direction and micro-nano structural unit lengths.

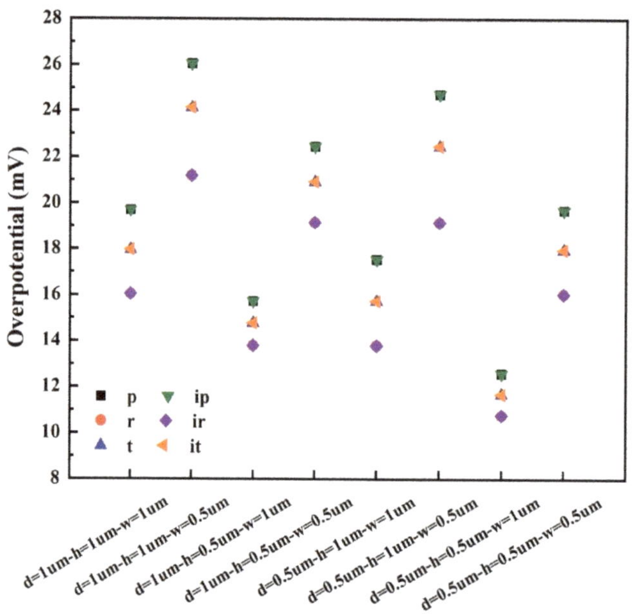

Figure 7. The overpotentials of micro-nano surface electrodes with different structure types and graphic parameters under the same current density of 10 mA cm^{-2}.

The overpotentials of micro-nano surface electrodes with different structure types and graphic parameters and the vertical flat electrode under the same current density of 10 mA cm^{-2} are shown in Table 2. The polarization curves of the micro-nano surface and vertical flat electrodes are shown in Figure 8. Figure 8a shows the polarization curves of ir in different graphic parameters, and Figure 8b shows the polarization curves of six structures at the graphic parameters of 0.5 μm, 0.5 μm, and 1 μm. It can be noted that the cell voltage and cathode overpotential of all micro-nano surface electrodes are smaller than those of the vertical flat electrode. Compared with the vertical flat electrode, for the same graphic parameters (0.5 μm, 0.5 μm, and 1 μm), the overpotential of p, r, t, ip, ir, and it decreased by 59.44%, 65.31%, 62.40%, 59.44%, 65.31%, and 62.40%, respectively; for the same structure, the overpotentials of ir with different graphic parameters are reduced by 48.29%, 31.69%, 55.54%, 38.27%, 55.54%, 38.27%, 65.31%, and 48.29%, respectively.

Table 2. The overpotentials of micro-nano surface electrodes with different structure types and graphic parameters and the vertical flat electrode under the same current density of 10 mA cm^{-2}.

Graphic Parameters (μm)			Overpotential (mV)						
Distance	Height	Width	p	r	t	ip	ir	it	flat-200
-	-	-	-	-	-	-		-	31.02
1	1	1	19.70	16.04	17.96	19.70	16.04	17.96	-
1	1	0.5	26.05	21.19	24.13	26.05	21.19	24.13	-
1	0.5	1	15.72	13.79	14.76	15.72	13.79	14.76	-
1	0.5	0.5	22.45	19.15	20.92	22.453	19.15	20.92	-
0.5	1	1	17.54	13.79	15.72	17.54	13.79	15.72	-
0.5	1	0.5	24.72	19.15	22.45	24.72	19.15	22.45	-
0.5	0.5	1	12.58	10.76	11.67	12.58	10.76	11.67	-
0.5	0.5	0.5	19.70	16.04	17.96	19.70	16.04	17.96	-

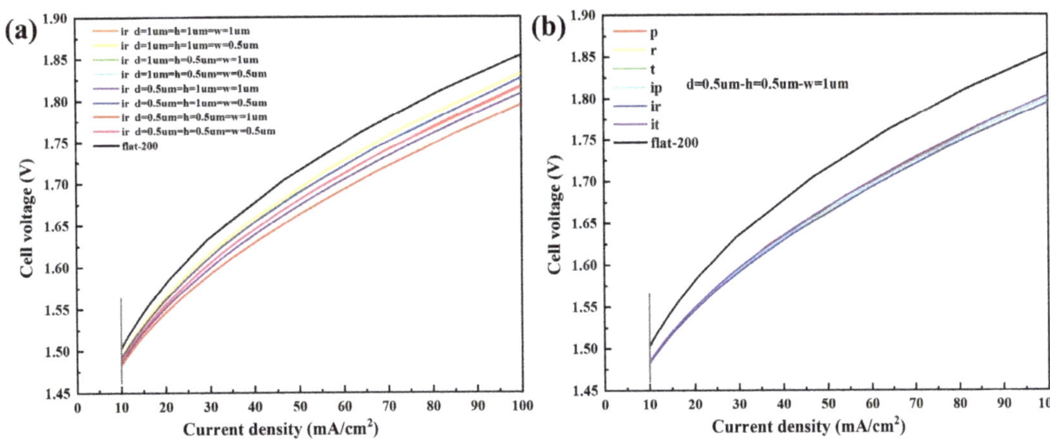

Figure 8. The polarization curves of the micro-nano surface and vertical flat electrodes. (**a**) The polarization curves of ir with different graphic parameters; (**b**) The polarization curves of six structures at the graphic parameters of 0.5 μm, 0.5 μm, and 1 μm.

The total length of micro-nano structural units and overpotential of each micro-nano surface electrode are shown in Table 3. The total length of micro-nano surface electrodes is larger than that of the vertical flat electrode. It can be concluded that the total length of micro-nano structural units affects the overpotential, and the larger the total length of micro-nano structural units, the lower the corresponding overpotential. Due to each group of structures (r and ir, p and ip, t and it) having the same total length of micro-nano structural units under the same graphic parameters, the overpotential is similar. Moreover, in these three groups of structures, the total length of micro-nano structural units corresponding to r and ir is more considerable, so the overpotential is the lowest. Due to the simulation model being a simplified 2D model of AWE, the length of micro-nano structural units is related to the contact area between the electrode and the electrolyte. Therefore, the increase in the length of micro-nano structural units means an increase in the contact area, which is helpful to increase the active site and reduce the cell voltage and overpotential, thus improving the electrolysis performance.

Table 3. The total length of micro-nano structural units and overpotential of each micro-nano surface electrode.

Graphic Parameters (μm)			Total Length of Micro-Nano Structure Units (mm)				Overpotential (mV)			
Distance	Height	Width	p/ip	r/ir	t/it	flat-200	p	r	t	flat-200
-	-	-	-	-	-	10	-		-	31.02
1	1	1	16.18	20.00	17.81	-	19.70	16.04	17.96	-
1	1	0.5	12.07	15.00	13.09	-	26.05	21.19	24.13	-
1	0.5	1	20.41	23.33	21.77	-	15.72	13.79	14.76	-
1	0.5	0.5	14.12	16.67	15.21	-	22.453	19.15	20.92	-
0.5	1	1	18.24	23.33	20.41	-	17.54	13.79	15.72	-
0.5	1	0.5	12.76	16.67	14.12	-	24.72	19.15	22.45	-
0.5	0.5	1	25.62	30.00	27.66	-	12.58	10.76	11.67	-
0.5	0.5	0.5	16.18	20.00	17.81	-	19.70	16.04	17.96	-

3.2.2. Void Fraction

The gas void fraction of micro-nano surface electrodes under different structure types and graphic parameters are shown in Table 4. The average void fraction produced on the electrode surface was selected as a comparison. It is seen that the average void fraction produced by all micro-nano surface electrodes is greater than that of the vertical flat electrode. The structures of r and ir, p and ip, and t and it have a similar void fraction under the same graphic parameters, and the void fraction of r and ir is the largest, followed by t and it. In addition, according to the order of the average void fraction from large to small, it is found that the corresponding graphic parameters of each structure are the same under the same ranking. Compared with the vertical flat electrode, for the same graphic parameters (0.5 μm, 0.5 μm, and 1 μm), the average void fraction of p, r, t, ir, ip, and it increases by 50.60%, 54.53%, 52.91%, 51.46%, 54.75%, and 53.28%, respectively; for the same structure, taking ir as an example, the average void fraction of each graphic parameters increases by 45.10%, 37.68%, 48.21%, 39.96%, 48.76%, 40.66%, 54.75%, and 45.17%, respectively. Obviously, the changing trend of the average void fraction is consistent with the overpotential. If the total length of micro-nano structural units is the same, the corresponding void fraction is similar; if the total length of micro-nano structural units is larger, the corresponding void fraction is larger. Therefore, it can be concluded that the total length of micro-nano structural units also affects the void fraction. Meanwhile, with the increase in the total length of micro-nano structural units, the catalytic area increases, which leads to the increase in void fraction.

Table 4. The average void fraction of micro-nano surface electrodes under different structures and graphic parameters and the vertical flat electrode at the same voltage (U = 2 V).

Graphic Parameters (μm)			Average Void Fraction (%)						
Distance	Height	Width	p	r	t	ip	ir	it	flat-200
-	-	-	-	-	-	-	-	-	2.15
1	1	1	3.00	3.11	3.06	3.02	3.12	3.06	-
1	1	0.5	2.84	2.97	2.90	2.85	2.96	2.90	-
1	0.5	1	3.12	3.20	3.16	3.12	3.19	3.14	-
1	0.5	0.5	2.93	3.02	2.97	2.94	3.01	2.97	-
0.5	1	1	3.06	3.19	3.13	3.08	3.20	3.13	-
0.5	1	0.5	2.88	3.00	2.94	2.88	3.02	2.93	-
0.5	0.5	1	3.24	3.32	3.29	3.26	3.33	3.29	-
0.5	0.5	0.5	3.00	3.12	3.06	3.02	3.12	3.06	-

Taking ir with distance, height, and width of 0.5 μm, 0.5 μm, and 1 μm as an example, the gas void fraction distribution of the micro-nano surface electrode and the vertical flat electrode is shown in Figure 9. It is seen that the gas distribution trend of the micro-nano surface electrode is consistent with that of the vertical flat electrode. However, the bubble layer produced by the micro-nano surface electrode is broader, and the local void fraction is higher.

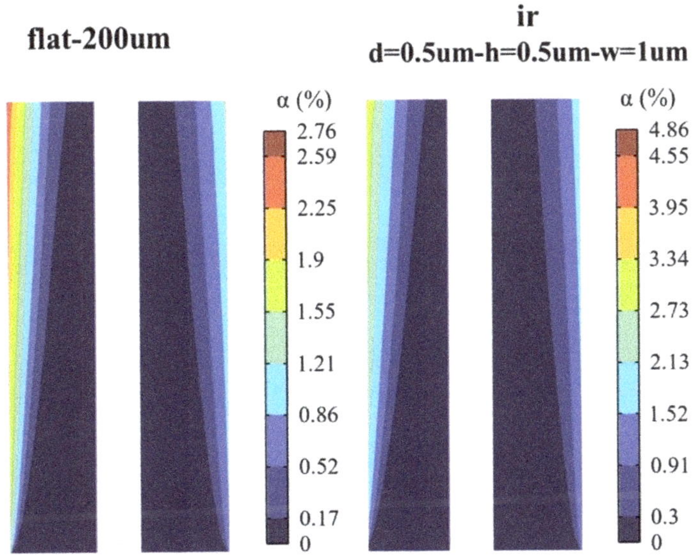

Figure 9. The gas void fraction distribution of micro-nano surface and vertical flat electrodes (U = 2 V).

The gas void fraction curves of the micro-nano surface electrode and the vertical flat electrode at the top of the H_2 compartment under the same voltage are shown in Figure 10. The gas void fraction curves of the micro-nano surface electrode and the vertical flat electrode under the same voltage are shown in Figure 11. The void fraction of the micro-nano surface electrode in the x and y directions is more significant than that of the vertical flat electrode. Compared with the vertical flat electrode, the micro-nano surface electrode can produce more hydrogen and increase the void fraction due to the accumulation of bubbles, which leads to more turbulent diffusion and promotes the lateral penetration of bubbles. Furthermore, due to the limitation of H_2 compartment height, bubbles cannot continue to diffuse upward and then accumulate, which leads to a significant increase in void fraction near the top of the micro-nano surface electrode compared with the vertical flat electrode.

Therefore, applying a micro-nano surface electrode helps increase hydrogen production and reduce cell voltage and overpotential, thus reducing energy consumption and improving the performance of the electrolyzer. Moreover, the effects of micro-nano surface electrodes with different structures and graphic parameters are different, so it is necessary to choose the appropriate structure type and graphic parameters according to the structural characteristics.

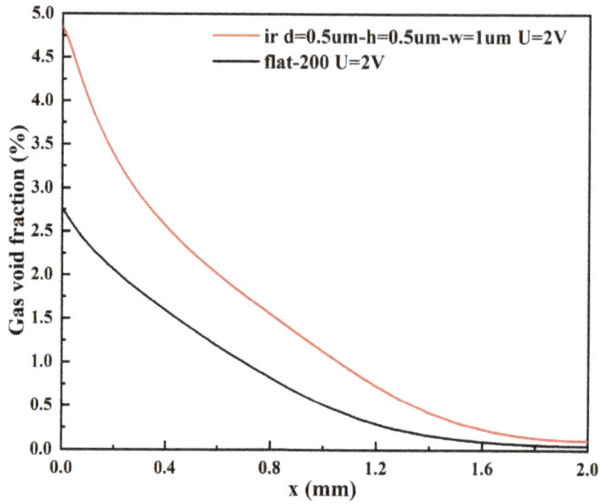

Figure 10. The gas void fraction curves of micro-nano surface and vertical flat electrodes at the top of the H_2 compartment under the same voltage (U = 2 V).

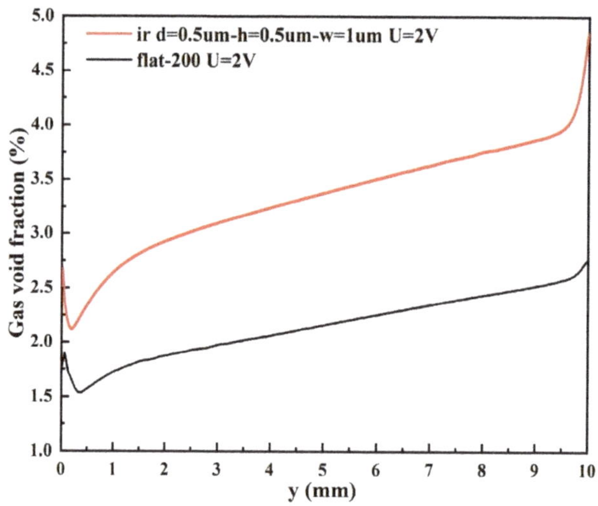

Figure 11. The gas void fraction curves of the micro-nano surface and vertical flat electrodes under the same voltage (U = 2 V).

4. Conclusions

In this study, the influence of micro-nano surface electrodes on the electrolytic performance of the AWE was numerically investigated. The results that can be inferred from this research are as follows,

- The micro-nano surface electrode can significantly reduce the cell voltage and cathode overpotential and increase the gas void fraction. Among the six structures and graphic parameters set in this paper, when the distance, height, and width of each structure are 0.5 μm, 0.5 μm, and 1 μm, respectively, the cell voltage and cathode overpotential are the lowest, and the void fraction is the highest, and rectangle/inverted rectangle has the best effect, followed by trapezoid/inverted trapezoid.

- The bubble layers produced by the micro-nano surface and vertical flat electrode are parabolic distributed. However, compared with the vertical flat electrode, the void fraction produced by the micro-nano surface electrode is higher, and the bubble layer is thicker at any voltage.
- The total length of micro-nano structural units affects the cell voltage, cathode overpotential, and void fraction. With the increase in the total length of micro-nano structural units, the catalytic area increases, the cell voltage and overpotential decrease, and the void fraction increases. Taking rectangular structural units with a distance, height, and width of 0.5 μm, 0.5 μm, and 1 μm, respectively, as an example, the total length of the corresponding micro-nano surface electrode is three times that of the vertical flat electrode. However, compared with the vertical flat electrode, the cathode overpotential decreases by 65.31% and the void fraction increases by 54.53%.

Supplementary Materials: The following supporting information can be downloaded at: https://www.mdpi.com/article/10.3390/ma15144927/s1, Table S1: Grid study of the present work for flat-200 um model. Figure S1: Mesh generation for the used geometry. Figure S2: Study of grid independence. Figure S3: Comparison between the model of this study and the experimental results. Reference [8] are cited in the supplementary materials

Author Contributions: Conceptualization, methodology, Y.X. and M.G.; software, Y.X.; resources, Y.X., M.G. and J.Y.; writing—original draft preparation, Y.X.; writing—review and editing, Y.X. and M.G.; supervision, Y.S., L.C. and S.M.; project administration, Y.S., L.C. and S.M.; funding acquisition, Y.S., L.C. and S.M. All authors have read and agreed to the published version of the manuscript.

Funding: This research was funded by the Basic Research Program of Qinghai Science and Technology Department, No. 2021-ZJ-948Q.

Institutional Review Board Statement: Not applicable.

Informed Consent Statement: Not applicable.

Data Availability Statement: Not applicable.

Conflicts of Interest: Authors do not have any conflict of interest.

Nomenclature

C_μ, $C_{\varepsilon,1}$, $C_{\varepsilon,2}$	constants in turbulence model, (-)	α_c	cathodic charge transfer coefficient, (-)
C_d	drag coefficient, (-)	α	void fraction, (-)
d_b	bubble diameter, (m)	μ_T	turbulent viscosity, (Pa s)
F	Faraday constant, (C mol^{-1})	μ	viscosity, (Pa s)
F_k	volume force, (N m^{-3})	ρ	density, (kg m^{-3})
g	gravitational acceleration, (m s^{-2})	η	overpotential, (V)
j	current density, (A m^{-2})	σ	electrolyte conductivity, (S m^{-1})
j_0	exchange current density, (A m^{-2})	ε	turbulent dissipation rate, (m^2 s^{-3})
k	turbulent kinetic energy, (m^2 s^{-2})	φ	electrical potential, (V)
M_g	molar mass of the gas (kg mol^{-1})	τ	stress tensor, (N m^{-2})
n	number of transferred electrons, (-)	Subscripts	
p	pressure, (Pa)	l	liquid phase
R	universal gas constant, (J mol^{-1} K^{-1})	g	gas phase
Re	Reynolds number, (-)	Abbreviations	
u	velocity, (m s^{-1})	AWE	alkaline water electrolyzer
T	temperature, (°C)	KOH	potassium hydroxide
x, y	spatial coordinates, (m)	HER	hydrogen evolution reaction
Greek symbols		OER	oxygen evolution reaction
α_a	anodic charge transfer coefficient, (-)	CFD	computational fluid dynamics

References

1. Luo, Y.; Zhang, Z.; Chhowalla, M.; Liu, B. Recent advances in design of electrocatalysts for high-current-density water splitting *Adv. Mater.* **2022**, *34*, 2108133. [CrossRef] [PubMed]
2. Darband, G.B.; Aliofkhazraei, M.; Shanmugam, S. Recent advances in methods and technologies for enhancing bubble detachment during electrochemical water splitting. *Renew. Sustain. Energy Rev.* **2019**, *114*, 109300. [CrossRef]
3. Wang, J.; Xu, F.; Jin, H.; Chen, Y.; Wang, Y. Non-noble metal-based carbon composites in hydrogen evolution reaction: Fundamentals to applications. *Adv. Mater.* **2017**, *29*, 1605838. [CrossRef] [PubMed]
4. Mei, B.; Barnoon, P.; Toghraie, D.; Su, C.-H.; Nguyen, H.C.; Khan, A. Energy, exergy, environmental and economic analyzes (4E) and multi-objective optimization of a PEM fuel cell equipped with coolant channels. *Renew. Sustain. Energy Rev.* **2022**, *157*, 112021. [CrossRef]
5. Burton, N.A.; Padilla, R.V.; Rose, A.; Habibullah, H. Increasing the efficiency of hydrogen production from solar powered water electrolysis. *Renew. Sustain. Energy Rev.* **2021**, *135*, 110255. [CrossRef]
6. Jia, Y.; Zeng, M.; Barnoon, P.; Toghraie, D. CFD simulation of time-dependent oxygen production in a manifold electrolyzer using a two-phase model. *Int. Commun. Heat Mass Transf.* **2021**, *126*, 105446. [CrossRef]
7. Zhang, H.; Maijenburg, A.W.; Li, X.; Schweizer, S.L.; Wehrspohn, R.B. Bifunctional heterostructured transition metal phosphides for efficient electrochemical water splitting. *Adv. Funct. Mater.* **2020**, *30*, 2003261. [CrossRef]
8. Morales-Guio, C.G.; Stern, L.A.; Hu, X. Nanostructured hydrotreating catalysts for electrochemical hydrogen evolution. *Chem. Soc. Rev.* **2014**, *43*, 6555–6569. [CrossRef]
9. Darband, G.B.; Aliofkhazraei, M.; Hyun, S.; Rouhaghdam, A.S.; Shanmugam, S. Electrodeposited Ni-Co-P hierarchical nanostructure as a cost-effective and durable electrocatalyst with superior activity for bifunctional water splitting. *J. Power Sources* **2019**, *429*, 156–167. [CrossRef]
10. Barati Darband, G.; Aliofkhazraei, M.; Sabour Rouhaghdam, A. Three-dimensional porous Ni-CNT composite nanocones as high performance electrocatalysts for hydrogen evolution reaction. *J. Electroanal. Chem.* **2018**, *829*, 194–207. [CrossRef]
11. Brown, D.E.; Mahmood, M.N.; Turner, A.K.; Hall, S.M.; Fogarty, P.O. Low overvoltage electrocatalysts for hydrogen evolving electrodes. *Int. J. Hydrogen Energy* **1982**, *7*, 405–410. [CrossRef]
12. Zhang, W.; Jia, Q.; Liang, H.; Cui, L.; Wei, D.; Liu, J. Iron doped Ni3S2 nanorods directly grown on FeNi3 foam as an efficient bifunctional catalyst for overall water splitting. *Chem. Eng. J.* **2020**, *396*, 125315. [CrossRef]
13. Wang, Y.; Qian, G.; Xu, Q.; Zhang, H.; Shen, F.; Luo, L.; Yin, S. Industrially promising IrNi-FeNi3 hybrid nanosheets for overall water splitting catalysis at large current density. *Appl. Catal. B Environ.* **2021**, *286*, 119881. [CrossRef]
14. Angulo, A.; van der Linde, P.; Gardeniers, H.; Modestino, M.; Fernández Rivas, D. Influence of bubbles on the energy conversion efficiency of electrochemical reactors. *Joule* **2020**, *4*, 555–579. [CrossRef]
15. Rodríguez, J.; Amores, E. CFD modeling and experimental validation of an alkaline water electrolysis cell for hydrogen production. *Processes* **2020**, *8*, 1634. [CrossRef]
16. Zarghamia, A.; Deena, N.G.; Vreman, A.W. CFD modeling of multiphase flow in an alkaline water electrolyzer. *Chem. Eng. Sci.* **2020**, *227*, 115926. [CrossRef]
17. Liu, L.; Cai, W.; Chen, Y.; Wang, Y. Fluid dynamics and mass transfer study of electrochemical oxidation by CFD prediction and experimental validation. *Ind. Eng. Chem. Res.* **2018**, *57*, 6493–6504. [CrossRef]
18. Stewart, K.; Lair, L.; De La Torre, B.; Phan, N.L.; Das, R.; Gonzalez, D.; Lo, R.C.; Yang, Y. Modeling and optimization of an alkaline water electrolysis for hydrogen production. In Proceedings of the 2021 IEEE Green Energy and Smart Systems Conference (IGESSC), Long Beach, CA, USA, 1–2 November 2021; pp. 1–6.
19. Wosiak, G.; Silva, J.D.; Sena, S.S.; Carneiro-Neto, E.B.; Lopes, M.C.; Pereira, E. Investigation of the influence of the void fraction on the energy consumption of a vertical electrolyser under natural convection. *J. Environ. Chem. Eng.* **2022**, *10*, 107577. [CrossRef]
20. Wosiak, G.; da Silva, J.; Sena, S.S.; de Andrade, R.N.; Pereira, E. CFD simulation and experimental comparison of bubble-induced convection during electrochemical water splitting. *Chem. Eng. J.* **2022**, *433*, 133194. [CrossRef]
21. Pu, Z.; Liu, T.; Amiinu, I.S.; Cheng, R.; Wang, P.; Zhang, C.; Ji, P.; Hu, W.; Liu, J.; Mu, S. Transition-metal phosphides: Activity origin, energy-related electrocatalysis applications, and synthetic strategies. *Adv. Funct. Mater.* **2020**, *30*, 2004009. [CrossRef]
22. Cao, J.; Zhou, J.; Zhang, Y.; Wang, Y.; Liu, X. Dominating role of aligned MoS2/Ni3S2 nanoarrays supported on three-dimensional Ni Foam with hydrophilic interface for highly enhanced hydrogen evolution reaction. *ACS Appl. Mater. Interfaces* **2018**, *10*, 1752–1760. [CrossRef]
23. Liu, C.; Gong, T.; Zhang, J.; Zheng, X.; Mao, J.; Liu, H.; Li, Y.; Hao, Q. Engineering Ni2P-NiSe2 heterostructure interface for highly efficient alkaline hydrogen evolution. *Appl. Catal. B Environ.* **2020**, *262*, 118245. [CrossRef]
24. Zhang, G.; Feng, Y.-S.; Lu, W.-T.; He, D.; Wang, C.-Y.; Li, Y.-K.; Wang, X.-Y.; Cao, F.-F. Enhanced catalysis of electrochemical overall water splitting in alkaline media by Fe doping in Ni3S2 nanosheet Arrays. *ACS Catal.* **2018**, *8*, 5431–5441. [CrossRef]
25. Qu, S.; Huang, J.; Yu, J.; Chen, G.; Hu, W.; Yin, M.; Zhang, R.; Chu, S.; Li, C. Ni3S2 nanosheet flowers decorated with CdS quantum dots as a highly active electrocatalysis electrode for synergistic water splitting. *ACS Appl. Mater. Interfaces* **2017**, *9*, 29660–29668. [CrossRef] [PubMed]
26. Huang, H.; Yan, M.; Yang, C.; He, H.; Jiang, Q.; Yang, L.; Lu, Z.; Sun, Z.; Xu, X.; Bando, Y.; et al. Graphene nanoarchitectonics: Recent advances in graphene-based electrocatalysts for hydrogen evolution reaction. *Adv. Mater.* **2019**, *31*, e1903415. [CrossRef] [PubMed]

7. Xia, Y.; Zhang, Z.; Gao, M.; Zheng, Y.; Chen, L.; Si, Y.; Mei, S. Enhancement in photoelectric performance of dye-sensitized solar cells with inverted pyramid structures based on nanoimprint lithography. *AIP Adv.* **2021**, *11*, 075103. [CrossRef]
8. Amores, E.; Ruiz, J.R.; Rodríguez, C.M.; Escribano, P.G. Study of an alkaline electrolyzer powered by renewable energy. In Proceedings of the COMSOL Conference, Stuttgart, Germany, 26–28 October 2011; COMSOL AB: Burlington, VT, USA, 2011.
9. Alkaline Electrolyzer. Available online: https://cn.comsol.com/model/download/981821/models.fce.alkaline_electrolyzer.pdf (accessed on 14 December 2021).
0. Hreiz, R.; Abdelouahed, L.; Fünfschilling, D.; Lapicque, F. Electrogenerated bubbles induced convection in narrow vertical cells: A review. *Chem. Eng. Res. Des.* **2015**, *100*, 268–281. [CrossRef]
1. Hreiz, R.; Abdelouahed, L.; Fünfschilling, D.; Lapicque, F. Electrogenerated bubbles induced convection in narrow vertical cells: PIV measurements and Euler–Lagrange CFD simulation. *Chem. Eng. Sci.* **2015**, *134*, 138–152. [CrossRef]
2. Rivera, F.F.; Pérez, T.; Castañeda, L.F.; Nava, J.L. Mathematical modeling and simulation of electrochemical reactors: A critical review. *Chem. Eng. Sci.* **2021**, *239*, 116622. [CrossRef]
3. Le Bideau, D.; Mandin, P.; Benbouzid, M.; Kim, M.; Sellier, M.; Ganci, F.; Inguanta, R. Eulerian two-fluid model of alkaline water electrolysis for hydrogen production. *Energies* **2020**, *13*, 3394. [CrossRef]
4. El-Askary, W.A.; Sakr, I.M.; Ibrahim, K.A.; Balabel, A. Hydrodynamics characteristics of hydrogen evolution process through electrolysis: Numerical and experimental studies. *Energy* **2015**, *90*, 722–737. [CrossRef]
5. Liu, C.L.; Sun, Z.; Lu, G.M.; Yu, J.G. Hydrodynamic characteristics of the two-phase flow field at gas-evolving electrodes: Numerical and experimental studies. *R. Soc. Open Sci.* **2018**, *5*, 171255. [CrossRef] [PubMed]
6. Boissonneau, P.; Byrne, P. An experimental investigation of bubble-induced free convection in a small electrochemical cell. *J. Appl. Electrochem.* **2000**, *30*, 767–775. [CrossRef]
7. Aldas, K.; Pehlivanoglu, N.; Mat, M. Numerical and experimental investigation of two-phase flow in an electrochemical cell. *Int. J. Hydrogen Energy* **2008**, *33*, 3668–3675. [CrossRef]
8. Le Bideau, D.; Mandin, P.; Benbouzid, M.; Kim, M.; Sellier, M. Review of necessary thermophysical properties and their sensivities with temperature and electrolyte mass fractions for alkaline water electrolysis multiphysics modelling. *Int. J. Hydrogen Energy* **2019**, *44*, 4553–4569. [CrossRef]
9. Mandin, P.; Hamburger, J.; Bessou, S.; Picard, G. Modelling and calculation of the current density distribution evolution at vertical gas-evolving electrodes. *Electrochim. Acta* **2005**, *51*, 1140–1156. [CrossRef]
0. *Electrochemistry Module User's Guide*; COMSOL Multiphysics®v6.0; COMSOL AB: Stockholm, Sweden, 2021.
1. Pfaffe, M. Which Current Distribution Interface Do I Use? Available online: https://www.comsol.com/blogs/current-distribution-interface-use/?setlang=1 (accessed on 10 February 2014).
2. Dickinson, E. Theory of Current Distribution. Available online: https://www.comsol.com/blogs/theory-current-distribution/?setlang=1 (accessed on 7 February 2014).
3. Mat, M.; Aldas, K. Application of a two-phase flow model for natural convection in an electrochemical cell. *Int. J. Hydrogen Energy* **2005**, *30*, 411–420. [CrossRef]
4. *CFD Module User's Guide*; COMSOL Multiphysics®v6.0; COMSOL AB: Stockholm, Sweden, 2020.
5. Le Bideau, D.; Chocron, O.; Mandin, P.; Kiener, P.; Benbouzid, M.; Sellier, M.; Kim, M.; Ganci, F.; Inguanta, R. Evolutionary design optimization of an alkaline water electrolysis cell for hydrogen production. *Appl. Sci.* **2020**, *10*, 8425. [CrossRef]
6. Marfaing, O.; Guingo, M.; Laviéville, J.; Bois, G.; Méchitoua, N.; Mérigoux, N.; Mimouni, S. An analytical relation for the void fraction distribution in a fully developed bubbly flow in a vertical pipe. *Chem. Eng. Sci.* **2016**, *152*, 579–585. [CrossRef]
7. Li, Y.; Zhang, H.; Xu, T.; Lu, Z.; Wu, X.; Wan, P.; Sun, X.; Jiang, L. Under-water superaerophobic pine-shaped Pt nanoarray electrode for ultrahigh-performance hydrogen evolution. *Adv. Funct. Mater.* **2015**, *25*, 1737–1744. [CrossRef]
8. Lu, Z.; Zhu, W.; Yu, X.; Zhang, H.; Li, Y.; Sun, X.; Wang, X.; Wang, H.; Wang, J.; Luo, J.; et al. Ultrahigh hydrogen evolution performance of under-water "superaerophobic" MoS(2) nanostructured electrodes. *Adv. Mater.* **2014**, *26*, 2683–2687. [CrossRef] [PubMed]

materials

MDPI

Article

Comparison of the Material Quality of $Al_xIn_{1-x}N$ (x—0–0.50) Films Deposited on Si(100) and Si(111) at Low Temperature by Reactive RF Sputtering

Michael Sun [1,*], Rodrigo Blasco [1,2], Julian Nwodo [3], María de la Mata [4], Sergio I. Molina [4], Akhil Ajay [5], Eva Monroy [5], Sirona Valdueza-Felip [1] and Fernando B. Naranjo [1]

1 Photonics Engineering Group, Electronics Department, University of Alcalá, Madrid-Barcelona Road km 33.6, 28871 Alcalá de Henares, Spain
2 Science, Computations and Technology Department, European University of Madrid, Tajo Street s/n, 28670 Villaviciosa de Odón, Spain
3 Fort Hare Institute of Technology, University of Fort Hare, Alice 5700, South Africa
4 Departamento Ciencia de los Materiales, I.M. y Q.I., IMEYMAT, Universidad de Cádiz, Campus Río San Pedro s/n, Puerto Real, 11510 Cádiz, Spain
5 CEA-Grenoble, INAC/PHELIQS, 17 av. des Martyrs, 38054 Grenoble, France
* Correspondence: michael.sun@uah.es

Abstract: $Al_xIn_{1-x}N$ ternary semiconductors have attracted much interest for application in photovoltaic devices. Here, we compare the material quality of $Al_xIn_{1-x}N$ layers deposited on Si with different crystallographic orientations, (100) and (111), via radio-frequency (RF) sputtering. To modulate their Al content, the Al RF power was varied from 0 to 225 W, whereas the In RF power and deposition temperature were fixed at 30 W and 300 °C, respectively. X-ray diffraction measurements reveal a c-axis-oriented wurtzite structure with no phase separation regardless of the Al content (x = 0–0.50), which increases with the Al power supply. The surface morphology of the $Al_xIn_{1-x}N$ layers improves with increasing Al content (the root-mean-square roughness decreases from ≈12 to 2.5 nm), and it is similar for samples grown on both Si substrates. The amorphous layer (~2.5 nm thick) found at the interface with the substrates explains the weak influence of their orientation on the properties of the $Al_xIn_{1-x}N$ films. Simultaneously grown $Al_xIn_{1-x}N$-on-sapphire samples point to a residual n-type carrier concentration in the 10^{20}–10^{21} cm^{-3} range. The optical band gap energy of these layers evolves from 1.75 to 2.56 eV with the increase in the Al. PL measurements of $Al_xIn_{1-x}N$ show a blue shift in the peak emission when adding the Al, as expected. We also observe an increase in the FWHM of the main peak and a decrease in the integrated emission with the Al content in room-temperature PL measurements. In general, the material quality of the $Al_xIn_{1-x}N$ films on Si is similar for both crystallographic orientations.

Keywords: AlInN; Si(100); Si(111); radio-frequency sputtering

1. Introduction

$Al_xIn_{1-x}N$ ternary semiconductor alloys have attracted huge interest for their application in solar cells, particularly after the revision of the indium nitride (InN) band gap energy in 2001 [1]. The direct band gap (i.e., high absorption coefficient) of $Al_xIn_{1-x}N$, tunable from the near-infrared (0.7 eV for InN [1]) to the ultraviolet (6.2 eV for AlN [2]) range, makes it an excellent candidate for developing photovoltaic devices in combination with silicon. In addition, this material shows high resistance to thermal and mechanical stress and irradiation with high-energy particles [3], which makes it suitable for space applications.

The synthesis of high-quality single-phase $AlxIn_{1-x}N$ layers is challenging due to the large difference in properties such as bonding energy, lattice constants, or growth temperature between the binary constituents, InN and AlN. The growth of $AlxIn_{1-x}N$ layers has been reported by various techniques, including metal–organic chemical vapor deposition (MOCVD) [4–7], molecular beam epitaxy (MBE) [8–12], elemental stacks annealing (ESA) [13,14], and reactive sputtering. Within this last technique, we can distinguish two approaches, one that uses a mixture of argon and nitrogen for the deposition [15–24] and another that uses only nitrogen [25–31], the latter being our case. Reactive sputtering also allows the deposition on large substrates and employs lower temperatures than MOCVD or MBE. However, the low-temperature deposition comes at the price of higher defect density. The presence of impurities such as hydrogen [32] and defects such as nitrogen vacancies [33] induces unintentional doping with a residual carrier concentration as high as 10^{21} cm^{-3}, which causes a blue shift in the optical band gap due to the Burstein–Moss effect [34].

$AlxIn_{1-x}N$ can be synthesized on different substrates, such as Si(111) [22,30,31,35–37], Si(100) [19,23,27,38], sapphire [17,22,26,30,35,37,38], glass [17,22,30,37], and GaAs [22]. However, the properties of the $Al_xIn_{1-x}N$ films strongly depend on the nature of the substrate. It is particularly interesting to study the deposition of silicon due to its potential for hybrid III-nitride/Si solar cells. Wurtzite III-nitrides are usually grown on silicon (111) due to the hexagonal symmetry of this crystallographic plane. However, today, silicon-based nanotechnology uses silicon (100) because of its lower amount of dangling bonds, which generate undesired recombination centers [34].

There are several studies about the growth of $Al_xIn_{1-x}N$ films on either Si(111) or Si(100) and its comparison with $Al_xIn_{1-x}N$ on sapphire substrates. Bashir et al. [35] deposited InN on Si(111) by RF sputtering and obtained large crystallite size, low microstrain, and low dislocation density. Afzal et al. [22] grew $Al_xIn_{1-x}N$ films on Si(111) at 300 °C using a magnetron cosputtering system and obtained polycrystalline films with the preferred orientation along the (101) direction, with higher crystallite size and lower surface roughness compared with other substrates such as GaAs and glass. However, the comparison of $Al_xIn_{1-x}N$ layers simultaneously grown on both Si(111) and Si(100) substrates by reactive RF sputtering has never been reported so far.

This work presents the study of the properties of $Al_xIn_{1-x}N$ layers with Al content ranging from 0% to 50% simultaneously deposited on silicon (100) and (111) via reactive RF sputtering at a relatively low substrate temperature, 300 °C. The layer characteristics in terms of structural, morphological, electrical, and optical properties are studied and compared considering both substrate orientations. Finally, we demonstrate the possibility of using silicon (100) as a feasible substrate for developing $Al_xIn_{1-x}N$ layers for device applications by taking advantage of its compatibility with today's silicon-based nanotechnology.

2. Materials and Methods

$Al_xIn_{1-x}N$ layers were simultaneously deposited in a reactive RF magnetron sputtering system (AJA International, ATC ORION-3-HV, Scituate, MA, USA) on three substrates: *p*-doped 375 μm thick Si(100), *p*-doped 500 μm thick Si(111) (both with a resistivity of 1–10 Ωcm), and on 500 μm thick (0001)-oriented sapphire. This system was equipped with a 2 inch confocal magnetron cathodes of pure In (4N5) and pure Al (5N). The base pressure of the system was in the order of 10^{-7} mbar. The substrate-target distance was fixed at 10.5 cm, and the temperature during the deposition was monitored with a thermocouple placed in direct contact with the substrate holder. The substrates were chemically cleaned in organic solvents before being loaded in the chamber, where they were outgassed for 30 min at 550 °C and then cooled down to the growth temperature. Prior to the deposition, the surface of the targets and the substrates were cleaned using a soft plasma etching with Ar (2 sccm and 20 W), causing no damage to the surface. $Al_xIn_{1-x}N$ layers were deposited in a pure N_2 atmosphere with a nitrogen flow of 14 sccm and a pressure of 0.47 Pa. The RF power applied to the Al target, P_{Al}, was set to 0, 100, 150, 175, and 225 W (samples M1–M5, respectively), while the RF power applied to the In target and the temperature were fixed to 30 W and 300 °C, respectively. A sputtering time of 3 h was used for the InN sample, 5 h for the sample with P_{Al} = 100 W, and 4 h for the rest. The thickness and deposition rate of the samples are summarized in Table 1.

Table 1. Summary of the deposition parameters and the structural and morphological analysis of $Al_xIn_{1-x}N$ on Si(100) and Si(111): c-axis parameter and Al mole fraction x extracted from HRXRD, layer thickness estimated from FESEM, and rms surface roughness measured by AFM.

Sample	Substrate	P_{Al} (W)	c (Å)	Al Mole Fraction x	FWHM Rocking Curve (°)	Thickness [1] (nm)	Deposition Rate [2] (nm/h)	Rms Surface Roughness [3] (nm)
M1	Si(100)	0	5.73	0	4.6	390	130	11.5
M2		100	5.61	0.12	2.4	790	160	9.5
M3		150	5.45	0.35	6.2	650	160	3.5
M4		175	5.42	0.40	3.2	620	155	3.5
M5		225	5.36	0.48	2.8	910	230	2.5
M1	Si(111)	0	5.73	0	4.7	380	125	13.0
M2		100	5.59	0.16	2.9	780	160	8.0
M3		150	5.44	0.36	6.1	640	160	3.5
M4		175	5.40	0.42	3.1	630	160	3.5
M5		225	5.35	0.49	2.8	585	150	2.5

[1] Standard error of ±30 nm. [2] Standard error of ±15 nm/h. [3] Standard error of ±0.3 nm.

The alloy mole fraction, crystalline orientation, and mosaicity of the films were evaluated by high-resolution X-ray diffraction (HRXRD) measurements using a PANalytical X'Pert Pro MRD system (Malvern, UK). In addition, the thicknesses of the layers were obtained from field-emission scanning electron microscopy (FESEM) images. Atomic force microscopy (AFM) was employed to study the surface morphology and estimate the root-mean-square (rms) surface roughness using a Bruker multimode Nanoscope IIIA microscope in tapping mode (Billerica, MA, USA). Additionally, transmission electron microscopy (TEM) provided a deeper understanding of the structural properties of the interface between the deposited material and the substrate. The electrical properties of the films were analyzed using room-temperature Hall-effect measurements in a conventional Van der Paw geometry.

Finally, photoluminescence measurements were carried out at room temperature by exciting the samples with ~20 mW of a continuous-wave laser diode emitting at λ = 405 nm focused on a 1 mm diameter spot. The emission was collected with a 193 mm focal-length Andor spectrograph equipped with a UV-extended silicon-based charge-coupled-device (CCD) camera operating at −65 °C between 200 and 1100 nm.

3. Results and Discussions

3.1. Structural Characterization

To study the structural quality of the layers, HRXRD $2\theta/\omega$ scans were carried out on the $Al_xIn_{1-x}N$ layers grown on Si(100) and Si(111), with the results shown in Figure 1a,b, respectively. All layers presented a wurtzite crystalline structure highly oriented along the c-axis, and no other crystallographic phases were detected. The increase in P_{Al} shifted the (0002) and (0004) reflection peaks assigned to $Al_xIn_{1-x}N$ toward higher diffraction angles, which confirmed the reduction in the c lattice parameter. The Al mole fraction of the alloy was estimated by applying Vegard's Law [39] to the AlN-InN system, using the c lattice parameter obtained from HRXRD and assuming fully relaxed layers. The calculated Al mole fraction, x, scales linearly with P_{Al} between x = 0 and x = 0.49 or 0.48, for Si(100) and Si(111) substrates, respectively, as summarized in Table 1.

The FWHM of the ω-scan (rocking curve) of the (0002) $Al_xIn_{1-x}N$ diffraction peak provides information about the mosaicity of the material. In this study, $Al_xIn_{1-x}N$ layers grown on both silicon substrates showed similar values, in the 3–6° range, without a clear trend (Table 1). This indicated that the mosaicity is independent of the crystal orientation of the silicon substrate.

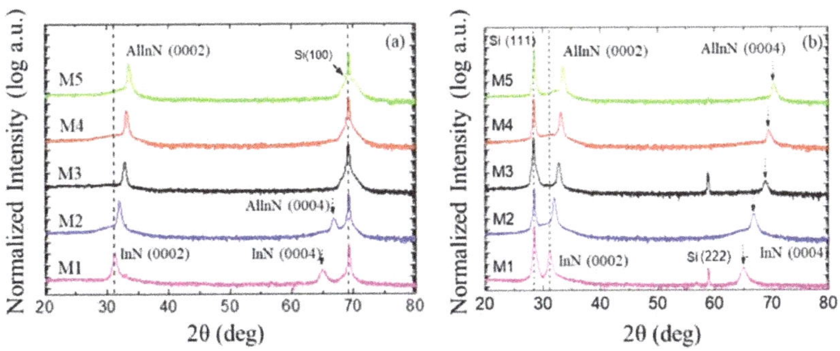

Figure 1. The $2\theta/\omega$ scans of the $Al_xIn_{1-x}N$ layers deposited on (**a**) Si(100) and (**b**) Si(111) for different P_{Al}. The only reflections assigned to $Al_xIn_{1-x}N$ were (0002) and (0004). The rest of the reflections were assigned to the substrates.

3.2. Morphological Characterization

In order to investigate the morphology of the layers, they were studied by FESEM and AFM techniques. Figure 2a–c show the FESEM images of samples grown on Si(100) and Si(111). The morphology of the layers evolved from nanocolumnar for pure InN (sample M1) toward grain-like compact when increasing the Al content (samples M3 and M5) for both substrate orientations. Such a trend was already observed in similar $Al_xIn_{1-x}N$ samples deposited on Si(111) by RF sputtering (40 W In, 300 °C) with similar Al compositions [31]. The observed phenomena could be attributed to changes in the surface diffusion of adatoms due to the increased kinetic energy of the incoming Al species, which can determine the layer morphology for both substrate orientations.

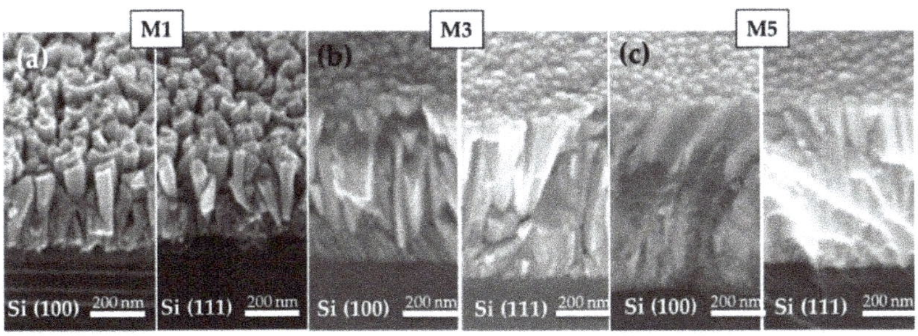

Figure 2. FESEM images of $Al_xIn_{1-x}N$ samples (**a**) M1, (**b**) M3, and (**c**) M5 on Si(100) (**left**) and Si(111) (**right**).

The observed morphological transition was accompanied by a modification of the sample surface roughness. The rms roughness was measured by AFM images scanned in a 2 × 2 μm area (Figure 3). The results showed a surface roughness evolution from 11.5 (Al content x = 0) to 2.5 nm (x ≈ 0.36) for Si(100) and from 13.0 (Al content x = 0) to 2.5 nm (x ≈ 0.36) for Si(111), as summarized in Table 1, and in agreement with previously published results [31]. The roughness remained almost constant for samples with an Al content in the range within x ≈ 0.36–0.42 (see Table 1), and it finally dropped up to ≈2.5 nm for an Al content of ≈50%. This surface roughness reduction was attributed to an increase in the adatom energy and mobility when increasing P_{Al}, in agreement with results obtained in similar $Al_xIn_{1-x}N$-on-Si(100) samples deposited at a higher temperature (550 °C) [27], where the surface roughness was 2.0 and 1.5 nm for x ≈ 0.35 and x ≈ 0.56, respectively.

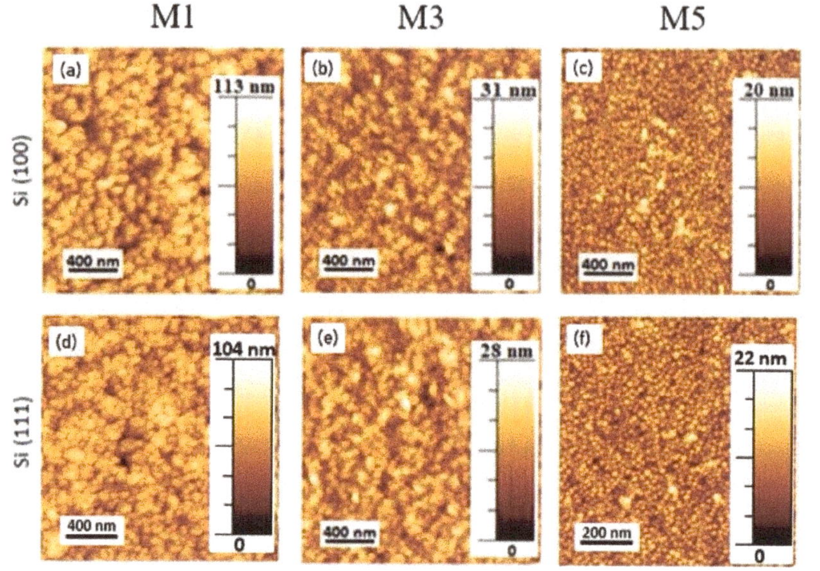

Figure 3. AFM images with a scanning area of 2 × 2 μm of InN and $Al_xIn_{1-x}N$ samples with P_{Al} = 0, 150, and 225 W grown on Si(100) (**a–c**) and Si(111) (**d–f**).

The interface between the $Al_xIn_{1-x}N$ and the silicon substrate was studied by transmission electron microscopy (TEM) measurements. Figure 4 shows the cross-sectional TEM images of an $Al_xIn_{1-x}N$ ($x \approx 0.36$) layer deposited on Si(100) and Si(111), evidencing the epitaxial growth along the c-axis for the two silicon orientations. In both cases, the images reveal the formation of an amorphous layer of ~2.5 nm at the layer/substrate interface (see the inset of both figures), which may have weakened the interactions between phases and reduced the influence of the silicon orientation on the quality of the nitride layer deposited on top. The similar structural quality obtained growing on both substrates was also confirmed by the comparable grain size estimated from STEM images (Figure 5a,b). Thus, the structural quality is conserved even when grown on a cubic substrate, although a clearer boundary between the amorphous interfacial layer and the nitride one was observed in this case.

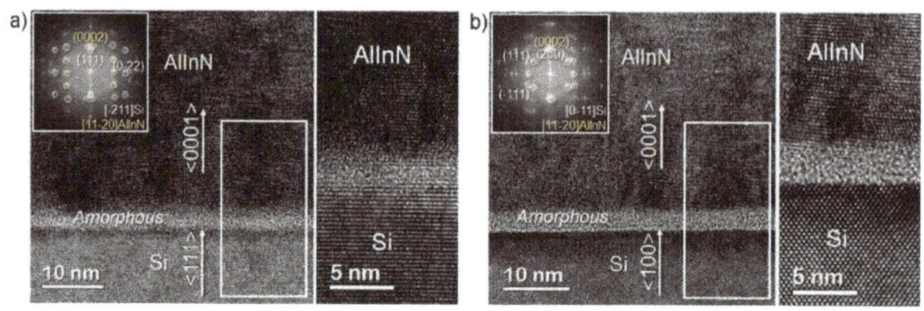

Figure 4. HRTEM images of $Al_{0.36}In_{0.64}N$ samples grown on (**a**) Si(100) and (**b**) Si(111), along with magnified details of the interphase (right side). Insets show the epitaxial relationship between the layer and substrate.

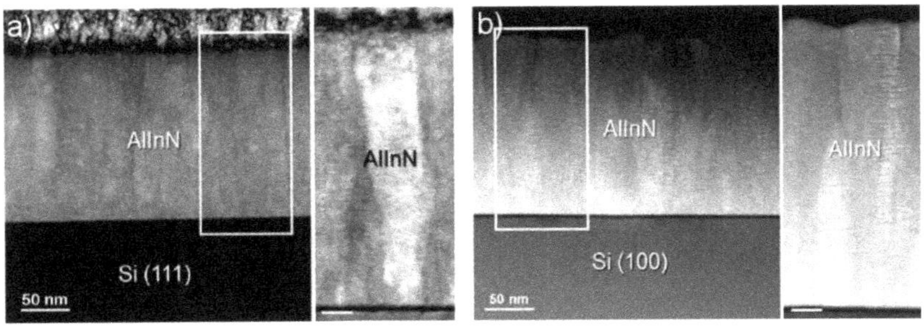

Figure 5. HRTEM images of $Al_{0.36}In_{0.64}N$ samples grown on (**a**) Si(100) and (**b**) Si(111) show a similar grain size on both substrates. Scale bar at the magnified details is 20 nm.

3.3. Electrical Characterization

The electrical properties of the $Al_xIn_{1-x}N$ layers could only be addressed for samples deposited on sapphire substrates, because the silicon conduction masked the layered signal whenever a silicon substrate was used. The values of resistivity, carrier concentration, and mobility were obtained for simultaneously grown layers with an Al content up to 0.32. Samples with higher Al content showed a resistivity above 10 mΩ·cm, making the Hall effect measurement unreliable.

The layer resistivity increased from 0.38 mΩ·cm for InN to 8 mΩ·cm for $Al_{0.32}In_{0.68}N$, while the carrier concentration decreased from 1.73×10^{21} cm^{-3} for InN to 2.48×10^{20} cm^{-3} for $Al_{0.32}In_{0.68}N$. On the other hand, the values of mobility showed no clear trend, starting with a value of 9.5 cm^2/V.s for InN and decreasing to 3.2 cm^2/V.s for $Al_{0.32}In_{0.68}N$ with a peak of 11.5 cm^2/V.s for $Al_{0.14}In_{0.86}N$. The values of resistivity and mobility obtained for the $Al_{0.32}In_{0.68}N$ sample are similar to those reported by Liu et al. [38] (1.2 mΩ·cm and 11.4 cm^2/V·s, respectively, for a ~90 nm $Al_{0.28}In_{0.72}N$ layer deposited by RF sputtering at 600 °C). The high carrier concentration of the layers is related to the unintentional doping from impurities such as hydrogen or oxygen during growth [32], and it was also observed by Nuñez-Cascajero et al. [26], where similar $Al_xIn_{1-x}N$ on sapphire with homogeneous distribution of oxygen were obtained.

3.4. Optical Characterization

The apparent optical band gap energy of the samples deposited on sapphire was estimated through room-temperature optical transmittance measurements following the procedure described in Ref. [26] (See Table 2 for all optical data). Figure 6 shows the squared absorption used for this estimation, obtained from the transmittance spectra depicted in the inset of the figure for each sample.

Table 2. Summary of the optical transmittance characterization at room temperature: apparent optical band gap energy (E_g^{Abs}), absorption band edge broadening (ΔE), and linear absorption well above the band gap (α_0) of the samples under study.

Sample	Al Mole Fraction x	α_0 ($\times 10^4$ cm^{-2})	E_g^{Abs}(eV) [1]	ΔE(meV) [2]
M1	0	17.2	1.70	160
M2	0.12	20.3	1.80	120
M3	0.35	18.4	2.10	210
M4	0.40	18.3	2.30	210
M5	0.48	10.0	2.60	180

[1] Standard error of ±0.03 eV. [2] Standard error of ±10 meV.

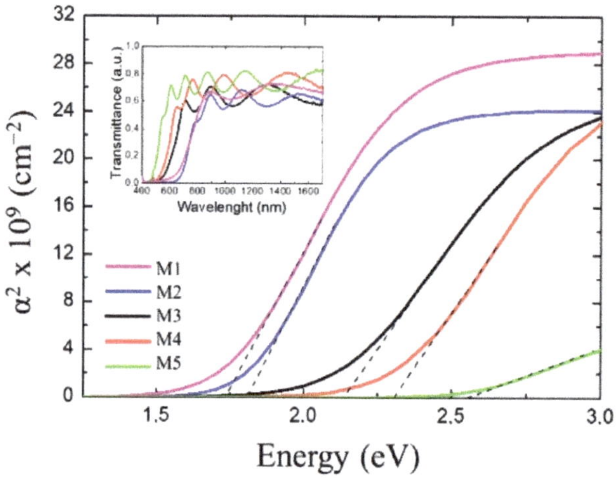

Figure 6. Squared absorption coefficient α^2 as a function of the energy extracted from the sigmoidal approximation of the $Al_xIn_{1-x}N$ layers grown on sapphire. Dashed lines are the linear fits used to estimate the apparent optical band gap energy of the samples $E_g{}^{Abs}$. Inset: transmittance spectra vs. wavelength of the same samples M1–M5.

As expected, the apparent optical band gap energy blue shifted with the Al mole fraction as following: Eg_{Abs} ~ 1.70 eV for InN (M1), 1.80 eV (M2), 2.10 eV (M3), 2.30 eV (M4), and 2.60 eV for $Al_{0.43}In_{0.57}N$ (M5). This blue shift in the optical band gap of the InN, compared with the theoretical of 0.7 eV, was attributed to the high residual carrier concentration of the layer.

Figure 7 shows the low-(11 K) and room-temperature (300K) PL emission of samples M1 (InN) and M2 ($Al_xIn_{1-x}N$, x—0.12, 0.16), grown on Si(100) and Si(111). No PL emission was observed for $Al_xIn_{1-x}N$ layers with higher Al content than 16%, independent of the crystal orientation of the substrate. The results obtained from the analysis of the PL measurements in terms of the main peak emission energy, FWHM, and integrated intensity are summarized in Table 3.

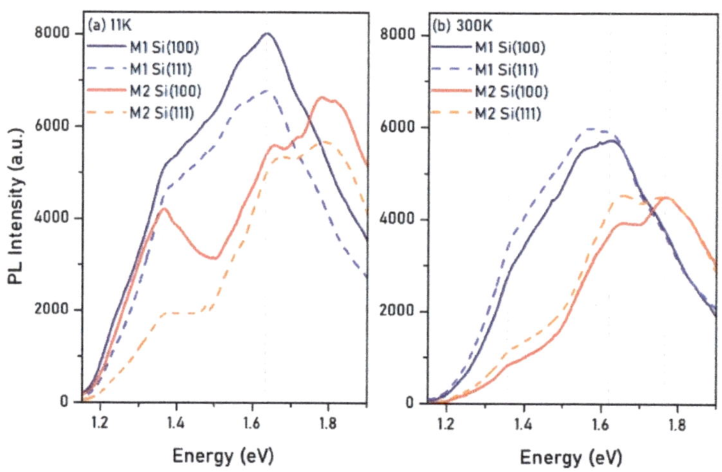

Figure 7. (**a**) Low-temperature (11K) and (**b**) room-temperature (300 K) PL emission of $Al_xIn_{1-x}N$ layers deposited on Si(100) and Si(111). For x > 0.16, no PL emission was observed.

Table 3. Summary of the analysis of the PL measurements at 11 K and 300 K of InN (M1) and $Al_xIn_{1-x}N$ (M2) on Si(100) and Si(111).

Sample	Temperature (K)	Substrate	Main Peak Emission Energy [1] (eV)	FWHM [2] (meV)	Integrated Intensity [3] (a.u.)
M1	11	Si(100)	1.60	560	4500
		Si(111)	1.60	515	3600
	300	Si(100)	1.60	460	2750
		Si(111)	1.60	465	2920
M2	11	Si(100)	1.80	565	3750
		Si(111)	1.80	480	3100
	300	Si(100)	1.80	500	2250
		Si(111)	1.75	490	2400

[1] Standard error of ±0.05 eV. [2] Standard error of ±5 meV. [3] Standard error of ±10.

The dominant room-temperature emission energy centered at ≈1.60 and ≈1.80 eV for the M1 (InN) and M2 ($Al_xIn_{1-x}N$, x—0.12, 0.16) samples deposited on both silicon substrates, respectively. The position of the emission energy practically stayed the same, while the intensity decreased when increasing the temperature from 11 to 300 K, as expected. However, the presence of an emission at room temperature was a clear indication of the good crystalline quality of the samples. The FWHM of the PL emission of the samples was similar for both types of substrates, being slightly higher for sample M2, probably due to the alloy disorder present in the $Al_xIn_{1-x}N$ layer.

Then, assuming that the band gap energy was similar for the samples grown on Si and sapphire, we could extract an approximate value for the Stokes shift as the difference between the band gap energy obtained from transmission measurements (Eg_{Abs}) and the PL emission energy (E_{PL}) at 300 K. The obtained Stokes shift was around ~130 and ~60 meV for InN (M1) and $Al_xIn_{1-x}N$ (M2), respectively. These values pointed to a reduced band tail for the $Al_xIn_{1-x}N$ samples compared with the InN ones, which could be related to the change in layer morphology (and probably the surrounding of the involved emission centers) when introducing aluminum into the InN binary.

Lastly, comparing each sample on both substrates, they showed a very similar emission shape and integrated intensity, even though the $Al_xIn_{1-x}N$ ones had double the layer thickness compared with their InN counterparts. This result pointed to an enhancement of the nonradiative recombination channels due to Al incorporation, which could increase the lattice disorder and defects.

4. Conclusions

$Al_xIn_{1-x}N$ films with low-to-mid Al content (*x*—0–0.50) were deposited via RF sputtering on different substrates, i.e., Si(100) and Si(111), for their comparison. The increase in the Al mole fraction improved the structural and morphological quality of the layers, achieving a minimum FWHM of the (0002) $Al_xIn_{1-x}N$ rocking curve of ~2.8° and a minimum rms surface roughness of ~2.5 nm for samples grown on both Si substrates with *x*—0.49. FESEM images showed a morphological transition from nanocolumnar toward a grain-like compact morphology when aluminum was introduced. Cross-sectional TEM images revealed a ~2.5 nm thick amorphous layer in the interface between the nitride material and the substrate, which could be responsible for the weak coupling between the active layer and the substrate. This finding allows the development of $Al_xIn_{1-x}N$ with similar material quality on both silicon substrate orientations.

Hall-effect measurements revealed a carrier concentration above 10^{20} cm^{-3} for the $Al_xIn_{1-x}N$ layers with $x < 0.32$, probably induced by the unintentional doping of the material during deposition. Additionally, the $Al_xIn_{1-x}N$ layers ($x \leq 0.16$) deposited on both Si substrate orientations exhibited similar PL emission in terms of shape, energy, FWHM, and integrated intensity at room temperature, showing a reduction in the PL emission efficiency when introducing the Al compared with the one obtained for InN layers.

In this work, we demonstrated the ability to produce high-quality $Al_xIn_{1-x}N$ layers on Si with low-to-mid Al content via RF sputtering regardless of the chosen substrate orientation.

Author Contributions: Investigation, R.B., J.N., M.d.l.M., S.I.M., A.A. and E.M.; Writing—original draft, M.S.; Writing—review & editing, S.V.-F. and F.B.N. All authors have read and agreed to the published version of the manuscript.

Funding: Partial financial support was provided by the projects: NERA (RTI2018-101037-B-I00), SINFOTON2-CM (P2018/NMT-4326), GRISA (CM/JIN/2021-021), and CAM-project (EPU-DPTO /2020/012).

Institutional Review Board Statement: Not applicable.

Informed Consent Statement: Not applicable.

Conflicts of Interest: The authors declare no conflict of interest.

References

1. Wu, J.; Walukiewicz, W.; Yu, K.M.; Ager, J.; Haller, E.E.; Lu, H.; Schaff, W.J.; Saito, Y.; Nanishi, Y. Unusual properties of th fundamental band gap of InN. *Appl. Phys. Lett.* **2002**, *80*, 3967–3969. [CrossRef]
2. Yim, W.M.; Stofko, E.J.; Zanzucchi, P.J.; Pankove, J.I.; Ettenberg, M.; Gilbert, S.L. Epitaxially grown AlN and its optical band ga *J. Appl. Phys.* **1973**, *44*, 292–296. [CrossRef]
3. Lu, N.; Ferguson, I. III-nitrides for energy production: Photovoltaic and thermoelectric applications. *Semicond. Sci. Technol.* **201** *28*, 074023. [CrossRef]
4. Hums, C.; Bläsing, J.; Dadgar, A.; Diez, A.; Hempel, T.; Christen, J.; Krost, A.; Lorenz, K.; Alves, E. Metal-organic vapor phas epitaxy and properties of AlInN in the whole compositional range. *Appl. Phys. Lett.* **2007**, *90*, 022105. [CrossRef]
5. Senda, S.; Jiang, H.; Egawa, T. AlInN-based ultraviolet photodiode grown by metal organic chemical vapor deposition. *App Phys. Lett.* **2008**, *92*, 203507. [CrossRef]
6. Yun, L.; Wei, T.; Yan, J.; Liu, Z.; Wang, J.; Li, J. MOCVD epitaxy of InAlN on different templates. *J. Semicond.* **2011** *32*, 093001. [CrossRef]
7. Dinh, D.V.; Hu, N.; Honda, Y.; Amano, H.; Pristovsek, M. Indium icorporation and optical properties of polar, semipolar an nonpolar InAlN. *Semicond. Sci. Technol.* **2019**, *35*, 035004. [CrossRef]
8. Wu, Y.-H.; Wong, Y.-Y.; Chen, W.-C.; Tsai, D.-S.; Peng, C.-Y.; Tian, J.-S.; Chang, L.; Chang, E.Y. Indium-rich InAlN films o GaN/sapphire by molecular beam epitaxy. *Mater. Res. Express* **2014**, *1*, 015904. [CrossRef]
9. Kamimura, J.; Kouno, T.; Ishizawa, S.; Kikuchi, A.; Kishino, K. Growth of high-In-content InAlN nanocolumns on Si (111) b RF-plasma-assisted molecular-beam epitaxy. *J. Cryst. Growth* **2007**, *300*, 160–163. [CrossRef]
10. Bhuiyan, A.G.; Islam, S.; Hashimoto, A. Molecular beam epitaxy of InAlN alloys in the whole compositional range. *AIP Adv* **2020**, *10*, 015053. [CrossRef]
11. Chen, W.-C.; Wu, Y.-H.; Tian, J.-S.; Yen, T.-C.; Lin, P.-Y.; Chen, J.-Y.; Hsiao, C.-N.; Chang, L. Effect of Growth Temperature o Structural Quality of In-Rich InxAl1−xN Alloys on Si (111) Substrate by RF-MOMBE. *ISRN Nanomater.* **2014**, *2014*, 1–6. [CrossRef
12. Yodo, T.; Ando, H.; Nosei, D.; Harada, Y. Growth and characterization of InN heteroepitaxial layers grown on Si substrates b ECR-assisted MBE. *Phys. Status Solidi Basic Res.* **2001**, *228*, 21–26. [CrossRef]
13. Afzal, N.; Devarajan, M.; Ibrahim, K. Growth of AlInN films via elemental layers annealing at different temperatures. *Mod. Phys Lett. B* **2015**, *29*, 1550169. [CrossRef]
14. Afzal, N.; Devarajan, M.; Subramani, S.; Ibrahim, K. Structural and surface analysis of AlInN thin films synthesized by elementa stacks annealing. *Mater. Res. Express* **2014**, *1*, 026403. [CrossRef]
15. Afzal, N.; Devarajan, M.; Ibrahim, K. Influence of substrate temperature on the growth and properties of reactively sputtere In-rich InAlN films. *J. Mater. Sci. Mater. Electron.* **2016**, *27*, 4281–4289. [CrossRef]
16. He, H.; Cao, Y.; Fu, R.; Guo, W.; Huang, Z.; Wang, M.; Huang, C.; Huang, J.; Wang, H. Band gap energy and bowing parameter o In-rich InAlN films grown by magnetron sputtering. *Appl. Surf. Sci.* **2010**, *256*, 1812–1816. [CrossRef]
17. Guo, Q.; Okazaki, Y.; Kume, Y.; Tanaka, T.; Nishio, M.; Ogawa, H. Reactive sputter deposition of AlInN thin films. *J. Cryst. Growt* **2007**, *300*, 151–154. [CrossRef]
18. Guo, Q.; Yahata, K.; Tanaka, T.; Nishio, M.; Ogawa, H. Growth and characterization of reactive sputtered AlInN films. *Phys Status Solidi* **2003**, 2533–2536. [CrossRef]
19. Lv, W.; Shen, L.; Liu, J.; Chen, J.; Wu, L.; Qi, D.; Zhang, G.; Li, X. Mechanical properties of single-phase Al1−In N films across th compositional range ($0 \leq x \leq 0.7$) grown by radio-frequency magnetron sputtering. *Appl. Surf. Sci.* **2019**, *504*, 144335. [CrossRef
20. Zhou, Y.; Peng, W.; Li, J.; Liu, Y.; Zhu, X.; Wei, J.; Wang, H.; Zhao, Y. Substrate temperature induced physical property variation o $In_xAl_{1-x}N$ alloys prepared on Al2O3 by magnetron sputtering. *Vacuum* **2020**, *179*, 109512. [CrossRef]

21. Zhao, Y.; Wang, H.; Li, X.; Li, J.; Shi, Z.; Wu, G.; Zhuang, S.; Yin, C.; Yang, F. Parametric study on the well-oriented growth of InxAl1-xN nanodots by magnetron sputtering. *Mater. Sci. Semicond. Process.* **2019**, *102*, 104583. [CrossRef]
22. Afzal, N.; Devarajan, M.; Ibrahim, K. A comparative study on the growth of InAlN films on different substrates. *Mater. Sci. Semicond. Process.* **2016**, *51*, 8–14. [CrossRef]
23. He, H.; Cao, Y.; Fu, R.; Wang, H.; Huang, J.; Huang, C.; Wang, M.; Deng, Z. Structure and optical properties of InN and InAlN films grown by rf magnetron sputtering. *J. Mater. Sci. Mater. Electron.* **2010**, *21*, 676–681. [CrossRef]
24. Afzal, N.; Devarajan, M.; Ibrahim, K. Synthesis of aluminium indium nitride (AlInN) thin films by stacked elemental layers method. *Eur. Phys. J. Appl. Phys.* **2014**, *67*, 10301. [CrossRef]
25. Valdueza-Felip, S.; Núñez-Cascajero, A.; Blasco, R.; Montero, D.; Grenet, L.; de la Mata, M.; Fernández, S.; Marcos, L.R.-D.; Molina, S.I.; Olea, J.; et al. Influence of the AlN interlayer thickness on the photovoltaic properties of in-rich AlInN on Si heterojunctions deposited by RF sputtering. *AIP Adv.* **2018**, *8*, 115315. [CrossRef]
26. Núñez-Cascajero, A.; Valdueza-Felip, S.; Monteagudo-Lerma, L.; Monroy, E.; Taylor-Shaw, E.; Martin, R.W.; González-Herráez, M.; Naranjo, F.B. In-rich $Al_xIn_{1-x}N$ grown by RF-sputtering on sapphire: From closely-packed columnar to high-surface quality compact layers. *J. Phys. D Appl. Phys.* **2017**, *50*, 065101. [CrossRef]
27. Blasco, R.; Valdueza-Felip, S.; Montero, D.; Sun, M.; Olea, J.; Naranjo, F.B. Low-to-Mid Al Content (x = 0–0.56) Al_xIn_{1-x} N Layers Deposited on Si(100) by Radio-Frequency Sputtering. *Phys. Status Solidi* **2020**, *257*, 1900575. [CrossRef]
28. Núñez-Cascajero, A.; Valdueza-Felip, S.; Blasco, R.; de la Mata, M.; Molina, S.I.; Gonzalez-Herraez, M.; Monroy, E.; Naranjo, F. Quality improvement of AlInN/p-Si heterojunctions with AlN buffer layer deposited by RF-sputtering. *J. Alloy. Compd.* **2018**, *769*, 824–830. [CrossRef]
29. Blasco, R.; Núñez-Cascajero, A.; Jiménez-Rodríguez, M.; Montero, D.; Grenet, L.; Olea, J.; Naranjo, F.B.; Valdueza-Felip, S. Influence of the AlInN Thickness on the Photovoltaic Characteristics of AlInN on Si Solar Cells Deposited by RF Sputtering. *Phys. Status Solidi* **2019**, *216*, 1800494. [CrossRef]
30. Núñez-Cascajero, A.; Blasco, R.; Naranjo, F.N.; Valdueza-Felip, S. High quality Al0.37In0.63N layers grown at low temperature (<300 °C) by radio-frequency sputtering. *Mater. Sci. Semicond. Process.* **2019**, *100*, 8–14. [CrossRef]
31. Núñez-cascajero, A.; Monteagudo-Lerma, L.; Valdueza-Felip, S.; Navío, C.; Monroy, E.; González-Herráez, M.; Naranjo, F.B. Study of high In-content AlInN deposition on p-Si (111) by RF-sputtering. *Jpn. J. Appl. Phys.* **2016**, *55*, 05FB07. [CrossRef]
32. Darakchieva, V.; Lorenz, K.; Barradas, N.P.; Alves, E.; Monemar, B.; Schubert, M.; Franco, N.; Hsiao, C.-L.; Chen, L.-C.; Schaff, W.J.; et al. Hydrogen in InN: A ubiquitous phenomenon in molecular beam epitaxy grown material. *Appl. Phys. Lett.* **2010**, *96*, 081907. [CrossRef]
33. Darakchieva, V.; Barradas, N.P.; Xie, M.-Y.; Lorenz, K.; Alves, E.; Schubert, M.; Persson, P.; Giuliani, F.; Munnik, F.; Hsiao, C.-L.; et al. Role of impurities and dislocations for the unintentional n-type conductivity in InN. *Phys. B Condens. Matter* **2009**, *404*, 4476–4481. [CrossRef]
34. Butcher, K.; Tansley, T. InN, latest development and a review of the band-gap controversy. *Superlattices Microstruct.* **2005**, *38*, 1–37. [CrossRef]
35. Bashir, U.; Hassan, Z.; Ahmed, N.M. A comparative study of InN growth on quartz, silicon, C-sapphire and bulk GaN substrates by RF magnetron sputtering. *J. Mater. Sci. Mater. Electron.* **2017**, *28*, 9228–9236. [CrossRef]
36. Han, Q.; Duan, C.; Du, G.; Shi, W.; Ji, L. Magnetron Sputter Epitaxy and Characterization of Wurtzite AlInN on Si(111) Substrates. *J. Electron. Mater.* **2010**, *39*, 489–493. [CrossRef]
37. Dong, C.J.; Xu, M.; Chen, Q.Y.; Liu, F.S.; Zhou, H.P.; Wei, Y.; Ji, H.X. Growth of well-oriented AlxIn1−xN films by sputtering at low temperature. *J. Alloy. Compd.* **2009**, *479*, 812–815. [CrossRef]
38. Liu, H.F.; Tan, C.C.; Dalapati, G.K.; Chi, D.Z. Magnetron-sputter deposition of high-indium-content *n*-AlInN thin film on *p*-Si(001) substrate for photovoltaic applications. *J. Appl. Phys.* **2012**, *112*, 063114. [CrossRef]
39. Von, L.V. Die Konstitution der Mischkristalle und die Raumfullung der Atome. *Z. Phys.* **1921**, *5*, 17–26.

materials

MDPI

Article

Effect of Argon on the Properties of Copper Nitride Fabricated by Magnetron Sputtering for the Next Generation of Solar Absorbers

C. A. Figueira [1,2], G. Del Rosario [3], D. Pugliese [4,5], M. I. Rodríguez-Tapiador [2] and S. Fernández [2,*]

[1] Department of Applied Mathematics, Materials Science and Engineering and Electronic Technology, Higher School of Experimental Sciences and Technology (ESCET), Rey Juan Carlos University (URJC), C/Tulipán s/n, 28933 Móstoles, Spain

[2] Energy Department, Energy, Environmental and Technological Research Centre (CIEMAT), Av. Complutense 40, 28040 Madrid, Spain

[3] Technology Support Center CAT, Rey Juan Carlos University, C/Tulipán s/n, 28933 Móstoles, Spain

[4] Department of Electronics and Telecommunications (DET), Politecnico di Torino, 10129 Turin, Italy

[5] Department of Applied Science and Technology (DISAT) and RU INSTM, Politecnico di Torino, 10129 Turin, Italy

* Correspondence: susanamaria.fernandez@cieamat.es; Tel.: +34-91-3466039

Abstract: Copper nitride, a metastable semiconductor material with high stability at room temperature, is attracting considerable attention as a potential next-generation earth-abundant thin-film solar absorber. Moreover, its non-toxicity makes it an interesting eco-friendly material. In this work, copper nitride films were fabricated using reactive radio frequency (RF) magnetron sputtering at room temperature, 50 W of RF power, and partial nitrogen pressures of 0.8 and 1.0 on glass and silicon substrates. The role of argon in both the microstructure and the optoelectronic properties of the films was investigated with the aim of achieving a low-cost absorber material with suitable properties to replace the conventional silicon in solar cells. The results showed a change in the preferential orientation from (100) to (111) planes when argon was introduced in the sputtering process. Additionally, no structural changes were observed in the films deposited in a pure nitrogen environment. Fourier transform infrared (FTIR) spectroscopy measurements confirmed the presence of Cu–N bonds, regardless of the gas environment used, and XPS indicated that the material was mainly N-rich. Finally, optical properties such as band gap energy and refractive index were assessed to establish the capability of this material as a solar absorber. The direct and indirect band gap energies were evaluated and found to be in the range of 1.70–1.90 eV and 1.05–1.65 eV, respectively, highlighting a slight blue shift when the films were deposited in the mixed gaseous environment as the total pressure increased.

Keywords: copper nitride; reactive RF magnetron sputtering; gaseous environment; solar absorber

1. Introduction

Copper nitride (Cu_3N) is attracting the attention of researchers due to its interesting properties, primarily its anti-ReO_3 structure and its bandgap energy (E_g) value. Its main attraction is that it is a non-toxic, metastable, and low-cost semiconducting material with earth-abundant elements and high stability at room temperature (RT). In addition, different studies have reported a wide range of band gap energy values, i.e., from 0.8 eV to 1.9 eV, which are easily achievable by controlling the deposition conditions [1,2]. The density functional theory (DFT) calculations, on the other hand, predict a relatively smaller E_g, ranging from 0.23 eV to 1.0 eV [3]. In general, the materials required for photovoltaic cells are semiconductors with a band gap between 1.1 eV and 1.7 eV [4]. These values are necessary to promote an electron from the valence band to the conduction band, thus generating an electric current. In the case of Cu_3N, the reported E_g values suggest that this material can be a potential solar absorber for the next generation of photovoltaic devices [5].

In this sense, Cu_3N has garnered interest in different application fields, such as optical storage media [6], tunnel junctions [7], solar energy conversion [8], and photovoltaics [9], due to its unique crystal structure and physicochemical properties.

Cu_3N exhibits a cubic anti-ReO_3 structure. The cell structure is shown in Figure 1.

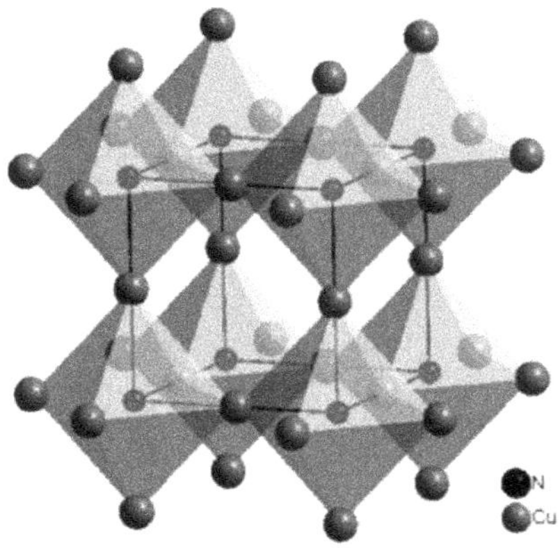

Figure 1. The cubic anti-ReO_3 crystal structure of Cu_3N [4].

The ideal crystal structure shows a cationic octahedral complex with an anion in the center, where the N atom occupies the Wyckoff position 1a (0, 0, 0) and the three Cu atoms are located at the 3d Wyckoff positions (1/2, 0, 0; 0, 1/2, 0; 0, 0, 1/2). In addition, in this structure, a void can be found at the body center (1/2, 1/2, 1/2) position [10]. The experimental lattice constant of cubic Cu_3N ranges from 0.3815 nm to 0.3885 nm [11]. Since Cu_3N is a non-stoichiometric interstitial compound, its lattice constant can be easily adjusted by varying the process conditions and/or doping with other atoms. Several studies have shown that the incorporation of transition metal atoms (e.g., Pd) generates a decrease in the energy gap value and increases the material's electrical conductivity [12,13].

Cu_3N films have been successfully fabricated using various methods, such as chemical vapor deposition (CVD) [14], pulsed laser deposition (PLD) [15], atomic layer deposition (ALD) [14], and reactive radio frequency (RF) magnetron sputtering [16]. Among these, sputtering deposition is highly preferable because it can be performed at low cost, without the use of toxic gases, and with low energy consumption because it can be carried out at RT.

Regarding the photovoltaic (PV) sector, Cu_3N has been extensively studied and has been proposed as a possible substitute for silicon, which is the most widely used material in solar panels. In addition, thanks to its good electrical properties and the ease with which it can be fabricated, the use of this material as a solar absorber could help avoid the high-temperature steps involved in the fabrication of solar-grade silicon. Furthermore, thanks to bipolar doping [17], this material can be profitably used in rectifying heterojunctions [18–20] as well as homojunctions [8–21], and in solar cells [8,17,22].

In this work, the impact of using argon (Ar) in the RF magnetron sputtering process on all the properties of the Cu_3N thin films fabricated on glass at RT have been thoroughly evaluated. In the material deposition, two different gaseous environments were used: an Ar-free environment based on pure nitrogen (N_2) and an environment consisting of a changeable mix of N_2 and Ar. Our main aim was to determine how the presence of Ar in the sputtering chamber might modify the films' properties. Within this framework, it is of paramount importance to establish the sputtering conditions under which the material could be considered more suitable to be used as a solar absorber in a photovoltaic device.

2. Materials and Methods

The Cu_3N thin films were deposited on silicon and glass substrates using reactive RF magnetron sputtering in a mono chamber (a commercial MVSystem). In this system, the RF-operated cathode was vertically adjustable, and the substrate was placed in front of the metallic target, a 99.99% Lesker Cu 3-inch diameter disk. The substrates had been prepared prior to the sputtering deposition. The silicon wafers were immersed in a hydrofluoric acid solution for 3 min and then ultrasonically cleaned for 3 min in a distilled water solution. The glass substrates were ultrasonically cleaned for 3 min in ethanol and distilled water solutions. The substrates were then dried by blowing nitrogen gas and loaded into the sputtering chamber. The base pressure of the chamber was around 2.6×10^{-5} Pa, and the distance between the target and the substrate was set at 98.8 mm. A pre-sputtering process to clean the target surface was carried out for 10 min using the sputtering conditions chosen for the deposition. The deposition time was set at 30 min and the films were deposited at RT and 50 W of RF power. The process gases used were N_2 (99.999%), with a flow rate of 20 sccm, and Ar (99.99995%), with a flow rate of 10 sccm, both controlled by MKS mass flow controllers (MFCs) (MKS Instruments, Andover, MA, USA). The partial pressure of the N_2, defined as $R = [N_2]/([N_2] + [Ar])$, was set at 0.8 and 1.0, and the total gas flow was maintained at 30 sccm. The total pressure in the chamber was varied from 3.5 Pa and 5.0 Pa by adjusting the position of the "butterfly" valve inside the magnetron system.

The film thickness was measured with a model Dektak 8 profilometer (Bruker, San José, CA, USA). For all samples, a tip force of 68.67 μN and a scan size of 2000 μm were used. The crystallinity of the Cu_3N films was determined by X-ray diffraction (XRD) using a commercial system (model PW3040/00 X' Pert MPD/MRD) (Malvern Panalytical Ltd., Malvern, UK) and Cu-kα radiation (λ = 0.15406 nm). The scanned 2θ range was 10–60°, with a step size of 0.01° and a time per step of 20 s. Ultra-high resolution field emission scanning electron microscope (FESEM) NovaNanoSEM 230, (FEI-company, Hillsboro, OR, USA) was used to determine the variation in the surface morphology as a function of the gaseous environment used. To enable the FESEM analysis, the samples had previously been covered with a thin platinum layer (2 nm). The FESEM images were taken at 30 kX, 60 kX, and 120 kX. This system was also equipped with an EDAX Apollo X silicon drift detector energy dispersive X-ray spectroscopy (EDS) system (EDAX Meter company, Leicester, UK). Several regions on the surfaces were analyzed at an acceleration voltage of 10 kV to qualitatively quantify the amount of Cu L_α, N K_α, and O K_α. Atomic force microscopy (AFM) (multimode nanoscope AFM model III, SPM, Veeco-Digital instrument Inc., Aschheim/Dornach Munich, Germany) was performed to investigate the surface of the specimens. The AFM measurements were carried out in tapping mode using silicon nitride AFM tips (OTR8, Veeco). The roughness was quantified by root mean square (RMS) in 1×1 μm^2 two-dimensional (2D) micrographs. The chemical composition of the Cu_3N thin films was characterized with Fourier transform infrared (FTIR) spectroscopy (Waltham, MA, USA) using a commercial Perkin Elmer Spectrum 100. The spectra were acquired using the transmittance mode (%) vs. the wave number, in the range of 400–4000 cm^{-1}. A surface analysis was carried out with X-ray photoelectron spectroscopy (XPS) using a PHI 5500 Multitechnique System (Physical Electronics® GmbH, Feldkirchen, Munich, Germany) and a monochromatic Al K-alpha X-ray source (1486.74 eV). This system was calibrated with the $3d_{5/2}$ line of Ag, perpendicular to the analyzer axis. The XPS spectra were analyzed with a Shirley background type and Gaussian–Lorentzian overlapping made with specific software. The data were fitted from the Cu-2p, N-1s, and O-1s states.

Finally, the optical transmittance was measured with a commercial Perkin Elmer Lambda 1050 spectrophotometer (Waltham, MA, USA) at normal incidence and RT and in the wavelength range of 400–2500 mm. The optical energy band gap was determined using the curve of $(\alpha h\nu)^p$ vs. photon energy, $h\nu$, by extrapolating the line to the abscissa of $h\nu$, according to the standard Tauc plot [23]. The index p, which characterizes the type of electronic transition, has specific values of 2 and 1/2 for the indirect and direct transitions, respectively. The refractive index was also evaluated in the near-infrared (NIR) region at the wavelengths of 825 nm, 1061 nm, 1312 nm, and 1533 nm, using the prism coupling technique (2010, Metricon Corporation, Pennington, NJ, USA). Ten scans were performed for each measurement and the estimated error of the measurement was ±0.001.

3. Results

Table 1 shows the data for the sputtering parameters used in the Cu_3N films deposition.

Table 1. Summary of the deposition conditions, the calculated deposition rate, and the measured thickness of the samples under study.

Sample	Total Pressure (Pa)	N_2 Flux (sccm)	Ar Flux (sccm)	N_2 Flow Ratio	Deposition Rate (nm/s)	Thickness (nm)
S1	5.0	20	10	0.8	0.05	96
S2	3.5	20	10	0.8	0.12	210
S3	5.0	20	–	1.0	0.06	115
S4	3.5	20	–	1.0	0.06	100

Two different behaviors were observed, depending on the process gas used. In the case of the sputtering deposition in the Ar + N_2 gas mixture, the rate decreased with the total pressure, as previously reported in the literature [17,24,25]. This effect can be ascribed to the increase in the number of collisions that the atoms in the plasma undergo when the total pressure rises. In this scenario, the mean free path of the atoms present in the plasma was clearly reduced, decreasing the deposition rate. On the other hand, the higher deposition rate observed in sample S2 was attributed to a poorer nitridation because of the lower pressure used and, hence, the deposition was performed in a more metallic regime [25,26].

In contrast, the deposition rate obtained when the sputtering process was performed in a pure N_2 environment revealed to be almost constant at the total pressure range used. Taking into account the fact that, in this case, the N atoms directly reacted with the Cu atoms from the target, the fact that no change in the deposition rate was obtained with the total pressure could indicate that a target "poisoning" effect might have occurred in the samples S3 and S4. This impeded the performance of the reactive magnetron sputtering, resulting in a significantly reduced sputtering yield and, hence, no increase in the deposition rate was observed [26].

Figure 2a shows the XRD patterns of the Cu_3N films deposited at the total pressures of 3.5 Pa and 5.0 Pa in a gas mixture of N_2 + Ar. As can be seen, the sample deposited at the total pressure of 5.0 Pa presented the (100) plane as the preferential orientation ($2\theta = 23.33°$), which is typical for N-rich thin films [19]. Another diffraction peak of lower intensity also appeared, corresponding to the (200) plane. In contrast, the sample deposited at the total pressure of 3.5 Pa presented the (111) plane ($2\theta = 40.79°$) as the preferential orientation, indicative of a Cu-rich phase [19]. In addition, other diffraction peaks, such as (100), (110), (200), and (210), appeared less intensely, and the appearance of the (100) plane, indicating an N-rich phase, suggests that the material continued to be non-stoichiometric [27]. According to reports in the literature [21], when sputtering deposition is carried out in a mixed gas atmosphere, the amount of N_2 at low pressure regimes is scarce, so Cu-rich (111) planes are more likely to form in the film. By increasing the total working pressure, the amount of N_2 in the chamber begins to increase slightly, and may be high enough to form N-rich (100) planes.

Figure 2b shows the XRD patterns of the Cu_3N films deposited at the total pressures of 3.5 Pa and 5.0 Pa in a pure N_2 atmosphere. Regardless of the total pressure, the patterns exhibited both the (100) and (200) diffraction planes of the Cu_3N structure, without the appearance of other planes [28]. This supports the idea of a possible target "poisoning" effect at total pressures higher than 3.5 Pa, which is in agreement with the absence of variation observed in the deposition rate for these samples [25,26].

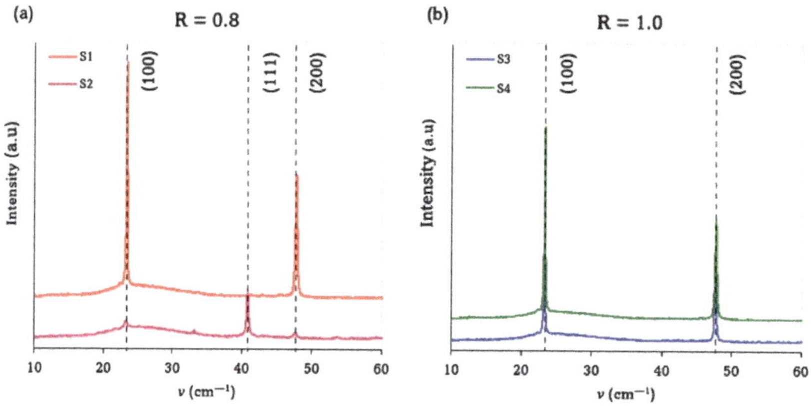

Figure 2. X-ray diffraction patterns of Cu_3N films deposited at RT and 50 W in (**a**) mixed Ar + N_2 and (**b**) pure N_2 environments.

Moreover, all the samples presented a cubic anti-ReO_3 structure, and the absence of impurity peaks (such as Cu or CuO_x) was indicative of a successful Cu nitridation to achieve Cu_3N thin films with a high degree of purity [28].

The lattice constant (1) was calculated according to the interplanar spacing using Bragg's law (2):

$$d_{hkl} = a/\sqrt{h^2 + k^2 + l^2} \tag{1}$$

$$d_{hkl} = \frac{n\lambda}{2\sin\theta}, \tag{2}$$

where d_{hkl} is the interplanar spacing of Miller indices h, k, and l, n is the order of diffraction, λ is the wavelength of X-rays, and θ is the diffraction angle. The theoretical value of the Cu_3N lattice constant used in the calculation is 0.3815 nm [5,27]. The averaged grain size was also determined by using Debye–Scherrer's Equation (3):

$$\tau = \frac{k\lambda}{\beta\cos\theta}, \tag{3}$$

where τ is the estimated grain size of the films, β is the full width at half maximum (FWHM) of the corresponding diffraction peak, θ is the diffraction angle, λ is the X-ray wavelength (0.15406 nm), and k is a constant.

Table 2 summarizes the parameters such as FWHM, lattice constant, and grain size calculated from the XRD patterns showed in Figure 2. The FWHM value obtained for the sample S2, deposited at the total pressure of 3.5 Pa in the Ar + N_2 environment, was significantly higher than that of the counterpart sample S4, deposited in a pure N_2 atmosphere. This could indicate a superior quality of the Cu_3N films when deposited without introducing Ar into the process gas. Apart from the sample S4, the lattice parameters obtained were larger than the theoretical ones, which could indicate an over-stoichiometry. Specifically, this state consisted of additional N atoms inserted into the Cu_3N lattice, probably as interstitials, revealing the formation of defects that could play an important role in increasing the lattice parameters [29]. This finding also agrees with the higher FWHM values obtained for the samples with larger lattice parameters. In all cases, these values are very close to that found in the literature for stoichiometric films fabricated by magnetron sputtering, i.e., 0.3815 nm [12,21,28–31].

Table 2. Main structural parameters derived from the XRD measurements.

Sample	Total Pressure (Pa)	N_2 Flow Ratio	Preferential Orientation	FWHM (°)	Lattice Constant (nm)	Grain Size (nm)
S1	5.0	0.8	(100)	0.1870	0.3817	42
S2	3.5	0.8	(111)	0.2755	0.3832	29
S3	5.0	1.0	(100)	0.1870	0.3820	42
S4	3.5	1.0	(100)	0.1771	0.3814	45

All the Cu_3N layers showed a single phase, with grain size values ranging from 29 nm to 45 nm, the grains of higher size being found in the samples with lower FWHM values. This would corroborate once again the better crystal quality of the samples deposited in the pure N_2 atmosphere.

The surface morphology of the samples was analyzed using AFM. Figure 3 depicts the 1×1 μm^2 3D micrographs of the sputtered films fabricated in the N_2 + Ar environment (Figure 3a) and in the pure N_2 atmosphere (Figure 3b).

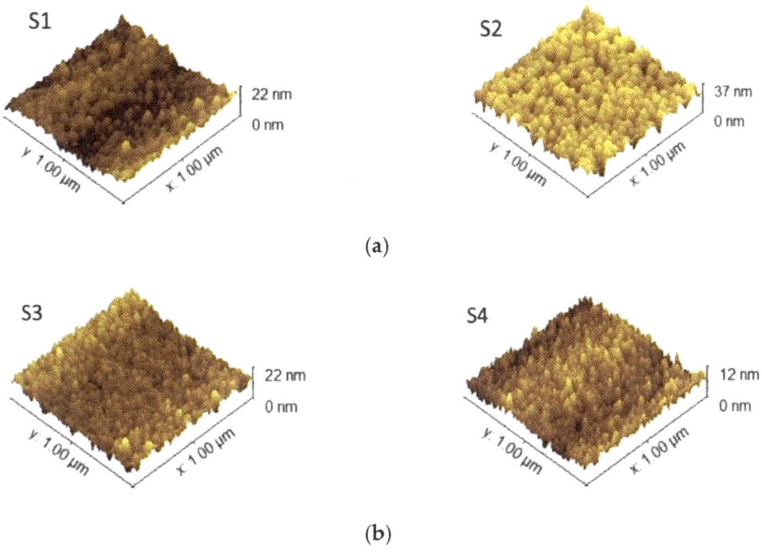

Figure 3. Three-dimensional micrographs (1×1 μm^2) of the Cu_3N films deposited in (**a**) N_2 + Ar mixture and (**b**) pure N_2 environments.

All the films presented a granular structure with tightly packed columnar grain features and sharp grain boundaries. The surfaces were smooth and uniform, regardless of the sputtering atmosphere used. However, two different morphologies were observed, depending on the composition of the process gas. This effect can be attributed to the increase in the deposition rate due to the introduction of Ar into the plasma, and also to the competition between the grain orientations present in the films, which led to changes in the morphology. Hence, the samples deposited in the N_2 + Ar environment (Figure 3a) showed a large number of void boundaries embedded in a complex pyramidal conical morphology, indicative of a competition between the (111) and (100) orientations, as revealed by the XRD patterns (Figure 2) [32]. The RMS values calculated for these samples were 3.03 nm (S1) and 5.12 nm (S2). A rougher surface was obtained for the sample S2 deposited at the total pressure of 3.5 Pa. Under such a pressure value, this sample would be deposited in a more metallic regime, where a lower number of N_2^+ ions were reacting with the sputtered

Cu atoms [25,26]. For this reason, its XRD pattern showed a competition between the grain orientations that led to a change in the preferred orientation, as was observed previously.

On the other hand, the films prepared in the pure N_2 atmosphere showed a relatively more compact surface, almost without the presence of voids. Under these conditions, more N_2^+ ions had the ability to combine with the Cu atoms, thus resulting in a high-density film with a smoother surface. In this case, the RMS values were very similar: 1.3 nm (S3) and 2.2 nm (S4). However, a slight reduction was observed when the sample was deposited at the highest deposition pressure used in this work. It is worth noting that the smoother surfaces exhibited by the films deposited at higher pressures was a measurable effect independent of the environment used and could be ascribed to the higher energy of the ion bombardment on the substrate [31]. Hence, it is clear how the presence of Ar in the process gas determined the type of surface morphology. The surface morphology and the roughness are parameters that must be taken into account when manufacturing a device.

Figure 4 presents plan-view FESEM images of the Cu_3N films deposited in the N_2 + Ar environment (Figure 4a) and in the pure N_2 atmosphere (Figure 4b).

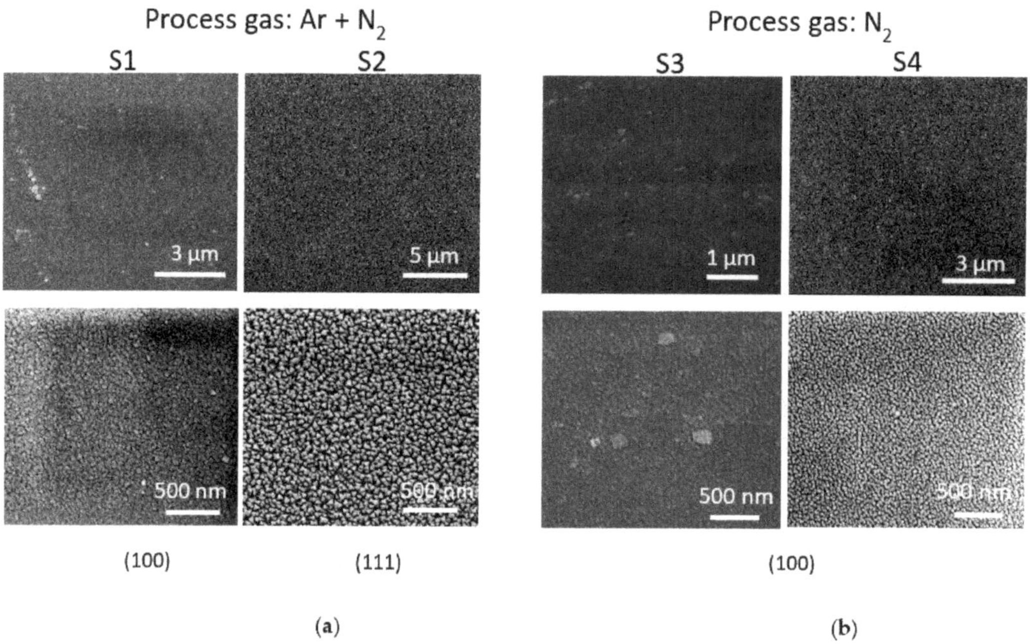

Figure 4. FESEM images of the surface of the Cu_3N samples deposited in (**a**) N_2 + Ar and (**b**) pure N_2 environments. Information concerning the preferred orientation extracted from the XRD analysis is also included.

These images confirmed that the surfaces were smooth and uniform, and mainly composed of columnar grains typical of the sputtering method [33]. Moreover, these results are consistent with those obtained from the previously reported AFM analysis. As can be observed, the grain shapes were influenced by the environment and by the total pressure used in the deposition. A pyramidal conical structure was observed in the films deposited at the lowest pressure of 3.5 Pa, and this was considerably oversized when the film was deposited in the N_2 + Ar environment. The shape of these grains turned out to be more spherical when the sample was deposited in a pure N_2 atmosphere. On the other hand, at the highest pressure of 5.0 Pa, the grains on the surface appeared to gain a nodular-like morphology, and this was once again oversized on the surface of the sample deposited in the mixed environment [32]. This morphology was characterized by the presence of a flat block of particles of uniform size, without visible cracks. These results are in line with the roughness calculated from the AFM images because these samples showed considerably smooth surfaces. The achievement of compact films would be beneficial for the production of heterojunctions needed for solar devices.

The EDS data, shown in Table 3, revealed a qualitative ratio of Cu to N lower than three and, hence, the obtained films did not become a stoichiometric material, being N-rich in all cases. From these measurements, the samples S1 and S2, deposited in the gas mixture, showed a very similar ratio, being within the confidence interval of less than 2%, whereas the sample S4, deposited at the pressure of 3.5 Pa, showed a lower ratio than S3, which was attributed to the lower presence of N in the plasma used to form the nitride film. However, the Cu/N ratios obtained for the samples deposited in the N_2 + Ar atmosphere were higher than those measured for the samples deposited in the pure N_2 environment. This can be explained by the higher number of N_2^+ ions present in the pure atmosphere in comparison with the mixed one at the same pressure. Finally, the presence of O on the surface was observed in all cases.

Table 3. Relative surface composition of the films under study, derived from the EDS analysis.

Sample	Total Pressure (Pa)	N_2 Flow Ratio	Cu (at%)	N (at%)	Cu/N	O (at%)
S1	5.0	0.8	44.29	20.38	2.17	35.34
S2	3.5	0.8	53.52	25.46	2.15	21.06
S3	5.0	1.0	44.49	21.55	2.06	33.97
S4	3.5	1.0	38.26	19.67	1.94	42.07

To confirm the chemical compositions of the Cu_3N films, X-ray photoelectron (XPS) measurements were carried out. The XPS spectra showed the photoelectron emissions from the Cu-2p, N-1s, and O-1s states. The data obtained are summarized in Table 4. The presence of O was also observed in this case, as in EDS measurements. The large amount of O detected can be explained by surface contaminations with C and O and, hence, by the appearance of C-C, C-O, and C=O bonds due to a long exposure to ambient air, as other authors have pointed out before [29]. It should be noted that the presence of O was not observed in the XRD spectra, which could confirm that the formation of such oxidized compounds would take place on the surface and that they would have an amorphous character. The extracted Cu/N ratio for the samples deposited in the pure N_2 environment was in the range of 1.46–1.56. In contrast, a slightly higher Cu/N ratio, ranging from 1.89 to 1.96, was obtained for the samples deposited in the N_2 + Ar mixture. All these values were significantly smaller than the expected ratio of 3 corresponding to a stoichiometric Cu_3N film. This would confirm the presence in all cases of the N-rich (100) plane in the XRD spectra, and also the increased incorporation of N in the samples deposited in the pure N_2 environment, as previously revealed by the EDS data.

Table 4. Relative surface composition of the films under study, derived from the XPS analysis.

Sample	Total Pressure (Pa)	N_2 Flow Ratio	Cu (at%)	N (at%)	Cu/N	O (at%)
S1	5.0	0.8	31.63	16.68	1.89	51.69
S2	3.5	0.8	34.05	17.34	1.96	48.61
S3	5.0	1.0	28.01	19.15	1.46	52.84
S4	3.5	1.0	27.63	17.65	1.56	54.71

To further investigate the molecular structure, a FTIR analysis was carried out. As is shown in Figure 5a,b, all the samples exhibited a single band at around 644–648 cm^{-1}, which confirmed the creation of the Cu–N bond. As other authors have reported, Cu_3N exhibits a prominent peak at 652 cm^{-1}, which can be ascribed to the intrinsic lattice mode vibration of Cu–N [30]. This would indicate that the amount of N was adequate to form the Cu_3N phase. A weak peak at around 819 cm^{-1}, assigned to the Cu–N_3 bond, was also observed in all cases. In addition, a peak at 2049 cm^{-1} appeared in the whole spectra pictured in Figure 5c,d, corresponding to the stretching vibration of N_3 azide and confirming the formation of an N-rich Cu_3N material, in agreement with the data shown previously.

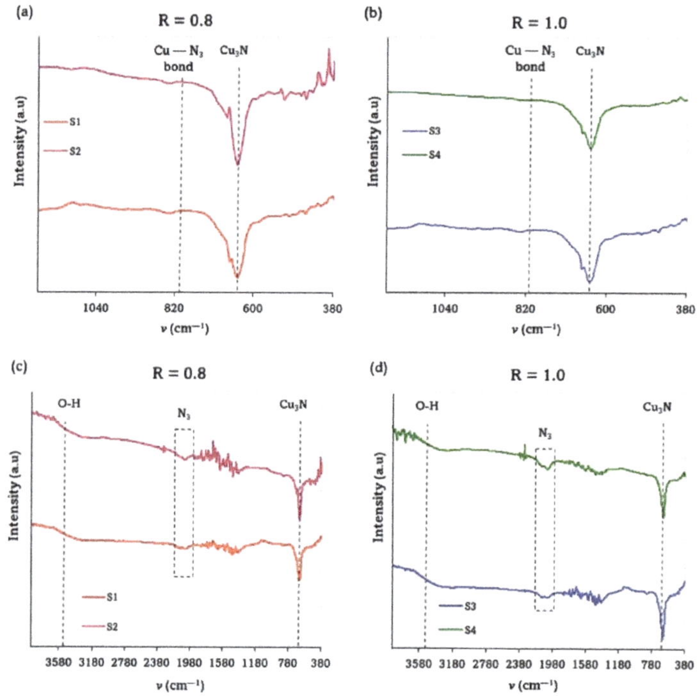

Figure 5. FTIR spectra of the Cu_3N films deposited in (**a**,**c**) an N_2 + Ar gas mixture and (**b**,**d**) a pure N_2 atmosphere.

From Figure 5a, it is possible to highlight that the samples S1 and S2 deposited in the N_2 + Ar environment presented a small shoulder at higher wavenumbers, i.e., at around 670 cm^{-1} and 690 cm^{-1}, respectively, and that this was almost undetectable in the samples deposited in the pure N_2 atmosphere. The presence of this additional band could be attributed to two effects: a possible oxidation of the surface sample (more probable in the samples S1 and S2, which showed voids in their morphology) and/or the formation of an incomplete Cu–N bond [34]. To confirm the presence of O in such high quantities as the EDS and XPS results suggest, the FTIR spectra were further analyzed at larger wavenumbers, as pictured in Figure 5c,d. As the whole spectra is revealed, the stretching vibrations of the hydroxyl group O-H can be observed in all the spectra in the broad region of 3775–3550 cm^{-1}, regardless of the gaseous environment [35].

Table 5 summarizes the FWHM values and the wavenumber at which the center of the prominent peak was located, as obtained from the FTIR spectra. Once more, the samples deposited in the pure N_2 atmosphere presented narrower peaks, and thus lower FWHM values, particularly the sample deposited at the lowest total pressure of 3.5 Pa, indicating its improved quality, in agreement with the extracted XRD results. In the case of the samples deposited in the N_2 + Ar mixture, considering the main peak and the small shoulder as a single peak, the FWHM value was much higher for the sample S2, indicating a preferred orientation of (111) and poorer crystal quality (see Table 2).

Table 5. Main parameters derived from the FTIR spectra. The FWHM values of the samples S1 and S2 have been estimated by considering the prominent peak and the small shoulder as a single peak.

Sample	Total Pressure (Pa)	N_2 Flow Ratio	FTIR Peak (cm^{-1})	FWHM (cm^{-1})
S1	5.0	0.8	644	62.2
S2	3.5	0.8	645	80.3
S3	5.0	1.0	644	60.2
S4	3.5	1.0	648	58.9

Finally, the effect of the sputtering environment on the optical properties of the samples was also evaluated. The band gap energy (E_g) was determined from the transmission spectra using the standard Tauc plot [23]. The direct band gap was calculated by plotting $(\alpha h\nu)^{1/2}$ vs. $h\nu$ and by extrapolating a full line to the abscissa of $h\nu$ (Figure 6a). Similarly, the indirect band gap was calculated by plotting $(\alpha h\nu)^2$ vs. $h\nu$ and extrapolating a full line to the abscissa (Figure 6b). The data obtained are summarized in Table 6.

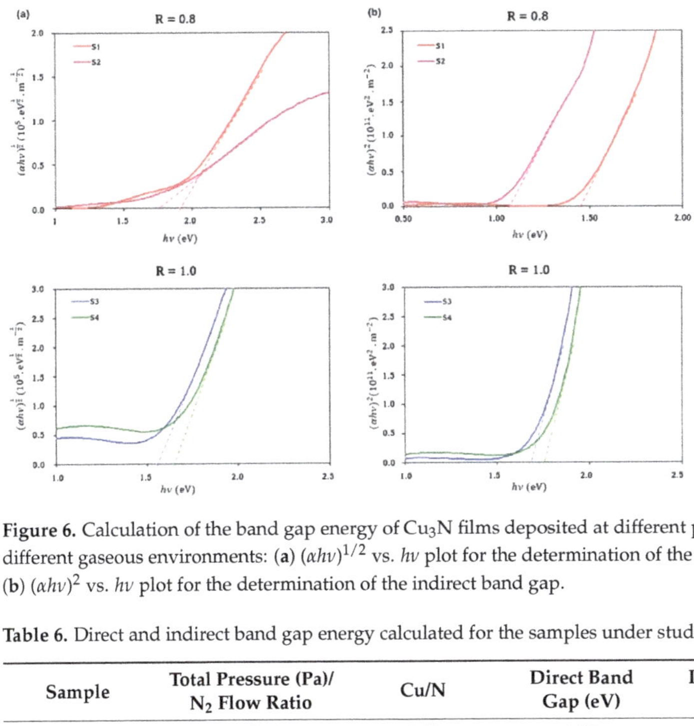

Figure 6. Calculation of the band gap energy of Cu_3N films deposited at different pressures and in different gaseous environments: (**a**) $(\alpha h\nu)^{1/2}$ vs. $h\nu$ plot for the determination of the direct band gap; (**b**) $(\alpha h\nu)^2$ vs. $h\nu$ plot for the determination of the indirect band gap.

Table 6. Direct and indirect band gap energy calculated for the samples under study.

Sample	Total Pressure (Pa)/ N_2 Flow Ratio	Cu/N	Direct Band Gap (eV)	Indirect Band Gap (eV)
S1	5.0/0.8	1.89	1.90	1.45
S2	3.5/0.8	1.96	1.77	1.05
S3	5.0/1.0	1.46	1.70	1.60
S4	3.5/1.0	1.56	1.79	1.65

The direct and indirect band gap values for the samples S1 and S2, deposited in the mixed N_2 + Ar atmosphere, were found to vary within the range of 1.77–1.90 eV and 1.05–1.45 eV, respectively, depending on the total pressure. As the work pressure increased, a blue shift was observed. Xiao et al. [36] attributed this fact to a slightly higher number of Cu atoms being unable to combine with N atoms to form Cu–N bonds due to the relatively low number of N atoms present in the gas mixture. Such Cu atoms would act as a light-scattering center, forming defects which would lead to a decrease in the band gap energy.

In this scenario, the energetic neutral N_2^+ ions in the plasma are dissociatively reflected by the target. This fact might account for a possible Cu re-sputtering, thus leading to a greater deficiency in the stoichiometric properties of Cu_3N films, as revealed by the EDS and XPS data.

On the other hand, the samples S3 and S4, deposited in the pure N_2 atmosphere, showed direct and indirect band gap values in the range of 1.70–1.79 and 1.60–1.65, respectively. In this case, the opposite phenomenon was observed, namely the band gap energy increased when the gas pressure decreased [28]. This behavior can be explained by the different nature of the gaseous environment in which the deposition was carried out. At a pressure of 3.5 Pa, an excess of Cu atoms may have continued forming the material due to the higher number of N atoms in the pure N_2 environment. However, when the N_2 pressure increased, the excess N that did not form Cu–N bonds could have integrated into the lattice in interstitial places, resulting in a worsening of the structural quality, as shown by the XRD results. This would lead to the appearance of defects and, hence, to a decrease in the band gap energy [16,37]. In any case, the band gap energy values obtained were in the range considered suitable for solar absorber applications [38].

Finally, the refractive indices of the Cu_3N films were measured at different wavelengths in the NIR region using the prism coupling technique. The obtained data are summarized in Table 7. As expected, the refractive indices showed a clear dependence on the measuring wavelength. Regardless of the gas environment used, the refractive index decreased with the wavelength, as reported in the literature [39].

Table 7. Refractive indices of the samples under study measured at different wavelengths in the NIR region.

Wavelength (nm)/$h\nu$ (eV)	S1	S2	S3	S4
825/1.503	1.521	—	1.522	1.512
1061/1.169	1.518	1.508	1.519	1.508
1312/0.945	1.515	1.506	1.515	1.506
1533/0.809	1.512	1.503	1.512	1.503

These values were relatively low compared with those found in the literature [5,39,40]. This could be attributed to two factors: (i) the possible low density of the films due to the packed columnar grain features observed in the AFM images combined with the N-rich nature indicated by the EDS and XPS data; and (ii) the high O presence on the surface, determined from EDS and XPS data, leading to the formation of a very thin amorphous CuO_x film (not observable in the XRD patterns).

4. Conclusions

In this work, Cu_3N films were prepared using reactive RF magnetron sputtering at RT at total pressures of 3.5 Pa and 5.0 Pa, and in different gaseous environments of N_2 and N_2 + Ar. The deposition rate calculated when the films were deposited in the mixture of N_2 and Ar gases increased with the total pressure, while a constant deposition rate of 0.06 nm/s was obtained when pure N_2 gas was used in the sputtering process, a fact attributed to a target "poisoning" effect. The XRD results revealed that the films had an anti-ReO_3 structure. The samples deposited in the pure N_2 environment showed the (100) plane as the preferential orientation. This plane corresponded to the N-rich phase of Cu_3N, indicative of an increased presence of N atoms rather than Cu atoms within the plasma. However, a transition from the (100) to (111) planes as the preferential orientation was highlighted when the total Ar + N_2 pressure decreased from 5.0 Pa to 3.5 Pa. At the same time, the FWHM values extracted from the XRD results revealed a worsening of the crystal quality when the films were deposited in the mixed gaseous environment. To conclude, a lattice constant larger than the theoretical one was obtained in any case, once again indicative of an N-rich phase. To support this hypothesis, FTIR, EDS, and XPS analyses were carried out, and the results confirmed the presence of an N-rich material and the formation of a Cu–N bond, regardless of the gaseous environment used. The XPS and EDS measurements also revealed a significant O contamination on the film surface, attributed to a long exposure in air. This contamination was also supported by the low refractive index values obtained for all of the films. Finally, the calculated optical band gap energies showed their strong dependence on the gaseous environment. In any case, a material with these values can be considered suitable for use as a solar absorber. However, all these results pointed to a preference for using (i) a pure N_2 environment and (ii) a range of moderate pressures, such as 3.5 Pa, due to the better crystal quality of the films deposited.

Author Contributions: Conceptualization, S.F.; methodology, C.A.F., M.I.R.-T. and S.F.; validation, C.A.F., G.D.R., D.P. and M.I.R.-T.; formal analysis, C.A.F., G.D.R. and M.I.R.-T.; investigation, D.P., M.I.R.-T. and S.F.; resources, G.D.R.; data curation, C.A.F., M.I.R.-T. and S.F.; writing—original draft preparation, C.A.F. and S.F.; writing—review and editing, D.P. and S.F.; visualization, M.I.R.-T.; supervision, S.F.; project administration, S.F.; funding acquisition, S.F. All authors have read and agreed to the published version of the manuscript.

Funding: This research was funded by MCIN/AEI/10.13039/501100011033, grant number PID2019-109215RB-C42. M.A. Rodríguez-Tapiador also acknowledges partial funding through MEDIDA C17.I2G: CIEMAT. Nuevas tecnologías renovables híbridas, Ministerio de Ciencia e Innovación, Componente 17 "Reforma Institucional y Fortalecimiento de las Capacidades del Sistema Nacional de Ciencia e Innovación". Medidas del plan de inversiones y reformas para la recuperación económica funded by the European Union—NextGenerationEU.

Institutional Review Board Statement: Not applicable.

Informed Consent Statement: Not applicable.

Data Availability Statement: Not applicable.

Acknowledgments: The authors would like to thank A. Soubrie from Centro de Microscopía Electrónica "Luis Bru" for her advice and AFM measurements.

Conflicts of Interest: The authors declare no conflict of interest.

References

1. Cho, S. Effect of substrate temperature on the properties of copper nitride thin films deposited by reactive magnetron sputtering. *Curr. Appl. Phys.* **2012**, *12*, S44–S47. [CrossRef]
2. Ji, A.L.; Huang, R.; Du, Y.; Li, C.R.; Wang, Y.Q.; Cao, Z.X. Growth of stoichiometric Cu_3N thin films by reactive magnetron sputtering. *J. Cryst. Growth* **2006**, *295*, 79–83. [CrossRef]
3. Rahmati, A.; Ghoohestani, M.; Badehian, H.; Baizaee, M. Ab. initio study of the structural, elastic, electronic and optical properties of Cu_3N. *Mater. Res.* **2014**, *17*, 303–310. [CrossRef]
4. Goetzberger, A.; Hebling, C. Photovoltaic materials, past, present, future. *Sol. Energy Mater. Sol. Cells* **2000**, *62*, 1–19. [CrossRef]
5. Mukhopadhyay, A.K.; Momin, M.A.; Roy, A.; Das, S.C.; Majumdar, A. Optical and electronic structural properties of Cu_3N thin films: A first-principles study (LDA + *U*). *ACS Omega* **2020**, *5*, 31918–31924. [CrossRef] [PubMed]
6. Zhao, Y.; Zhao, J.; Yang, T.; Zhang, J.; Yang, J.; Li, X. Enhanced write-once optical storage capacity of Cu_3N film by coupling with an Al_2O_3 protective layer. *Ceram. Int.* **2016**, *42*, 4486–4490. [CrossRef]
7. Borsa, D.M.; Grachev, S.; Presura, C.; Boerma, D.O. Growth and properties of Cu_3N films and Cu_3N/γ'-Fe_4N bilayers. *Appl. Phys. Lett.* **2002**, *80*, 1823–1825. [CrossRef]
8. Zakutayev, A.; Caskey, C.M.; Fioretti, A.N.; Ginley, D.S.; Vidal, J.; Stevanovic, V.; Tea, E.; Lany, S. Defect tolerant semiconductors for solar energy conversion. *J. Phys. Chem. Lett.* **2014**, *5*, 1117–1125. [CrossRef]
9. Ebaid, M.; Larson, D.M.; Bustillo, K.C.; Turner, J.H.; Cooper, J.K. Saw-tooth heat-cycling nitridation of metallic Cu yields first photoactive p-Cu_3N for PEC applications. *ACS Appl. Energy Mater.* **2020**, *3*, 10714–10721. [CrossRef]
10. Sahoo, G. Manipulating elastic and mechanical properties of Cu_3N through site selective Ag doping: First principles investigation. *Mater. Today Commun.* **2021**, *29*, 102950. [CrossRef]
11. Ji, A.; Yun, D.; Gao, L.; Cao, Z. Crystalline thin films of stoichiometric Cu_3N and intercalated Cu_3NM_x (M = metals): Growth and physical properties. *Phys. Status Solidi A* **2010**, *207*, 2769–2780. [CrossRef]
12. Ji, A.L.; Lu, N.P.; Gao, L.; Zhang, W.B.; Liao, L.G.; Cao, Z.X. Electrical properties and thermal stability of Pd-doped copper nitride films. *J. Appl. Phys.* **2013**, *113*, 043705. [CrossRef]
13. Gao, L.; Ji, A.L.; Zhang, W.B.; Cao, Z.X. Insertion of Zn atoms into Cu_3N lattice: Structural distortion and modification of electronic properties. *J. Cryst. Growth* **2011**, *321*, 157–161. [CrossRef]

14. Modin, A.; Kvashnina, K.O.; Butorin, S.M.; Werme, L.; Nordgren, J.; Arapan, S.; Ahuja, R.; Fallberg, A.; Ottosson, M. Electronic structure of Cu_3N films studied by soft X-ray spectroscopy. *J. Phys. Condens. Matter* **2008**, *20*, 235212. [CrossRef]
15. Gallardo-Vega, C.; De la Cruz, W. Study of the structure and electrical properties of the copper nitride thin films deposited by pulsed laser deposition. *Appl. Surf. Sci.* **2006**, *252*, 8001–8004. [CrossRef]
16. Nosaka, T.; Yoshitake, M.; Okamoto, A.; Ogawa, S.; Nakayama, Y. Copper nitride thin films prepared by reactive radio-frequency magnetron sputtering. *Thin Solid Films* **1999**, *348*, 8–13. [CrossRef]
17. Fioretti, A.N.; Schwartz, C.P.; Vinson, J.; Nordlund, D.; Prendergast, D.; Tamboli, A.C.; Caskey, C.M.; Tuomisto, F.; Linez, F.; Christensen, S.T.; et al. Understanding and control of bipolar self-doping in copper nitride. *J. Appl. Phys.* **2016**, *119*, 181508. [CrossRef]
18. Zhu, L.; Cao, X.; Gong, C.; Jiang, A.; Cheng, Y.; Xiao, J. Preparation of Cu_3N/MoS_2 heterojunction through magnetron sputtering and investigation of its structure and optical performance. *Materials* **2020**, *13*, 1873. [CrossRef]
19. Yu, A.; Hu, R.; Liu, W.; Zhang, R.; Zhang, J.; Pu, Y.; Chu, L.; Yang, J.; Li, X. Preparation and characterization of Mn doped copper nitride films with high photocurrent response. *Curr. Appl. Phys.* **2018**, *18*, 1306–1312. [CrossRef]
20. Tilemachou, A.; Zervos, M.; Othonos, A.; Pavloudis, T.; Kioseoglou, J. P-type iodine-doping of Cu_3N and its conversion to γ-CuI for the fabrication of γ-CuI/Cu_3N p-n heterojunctions. *Electron. Mater.* **2022**, *3*, 2. [CrossRef]
21. Chen, S.-C.; Huang, S.-Y.; Sakalley, S.; Paliwal, A.; Chen, Y.-H.; Liao, M.-H.; Sun, H.; Biring, S. Optoelectronic properties of Cu_3N thin films deposited by reactive magnetron sputtering and its diode rectification characteristics. *J. Alloys Comp.* **2019**, *789*, 428–434. [CrossRef]
22. Zakutayev, A. Design of nitride semiconductors for solar energy conversion. *J. Mater. Chem. A* **2016**, *4*, 6742–6754. [CrossRef]
23. Meymian, M.R.Z.; Heravi, A.D.; Mehr, A.K. Influence of bias voltage on optical and structural characteristics of Cu_3N films deposited by reactive RF magnetron sputtering in a pure nitrogen atmosphere. *Mater. Sci. Semicond. Process.* **2020**, *112*, 104995. [CrossRef]
24. Yue, G.H.; Yan, P.X.; Liu, J.Z.; Wang, M.X.; Li, M.; Yuan, X.M. Copper nitride thin film prepared by reactive radio-frequency magnetron sputtering. *J. Appl. Phys.* **2005**, *98*, 103506. [CrossRef]
25. Waite, M.M.; Shah, S.I. Target poisoning during reactive sputtering of silicon with oxygen and nitrogen. *Mater. Sci. Eng. B* **2007**, *140*, 64–68. [CrossRef]
26. Güttler, D.; Abendroth, B.; Grötzschel, R.; Möller, W.; Depla, D. Mechanisms of target poisoning during magnetron sputtering as investigated by real-time *in situ* analysis and collisional computer simulation. *Appl. Phys. Lett.* **2004**, *85*, 6134–6136. [CrossRef]
27. Gordillo, N.; Gonzalez-Arrabal, R.; Álvarez-Herrero, A.; Agulló-López, F. Free-carrier contribution to the optical response of N-rich Cu_3N thin films. *J. Phys. D Appl. Phys.* **2009**, *42*, 165101. [CrossRef]
28. Wang, J.; Chen, J.T.; Yuan, X.M.; Wu, Z.G.; Miao, B.B.; Yan, P.X. Copper nitride (Cu_3N) thin films deposited by RF magnetron sputtering. *J. Cryst. Growth* **2006**, *286*, 407–412. [CrossRef]
29. Majumdar, A.; Drache, S.; Wulff, H.; Mukhopadhyay, A.K.; Bhattacharyya, S.; Helm, C.A.; Hippler, R. Strain effects by surface oxidation of Cu_3N thin films deposited by DC magnetron sputtering. *Coatings* **2017**, *7*, 64. [CrossRef]
30. Du, Y.; Ji, A.L.; Ma, L.B.; Wang, Y.Q.; Cao, Z.X. Electrical conductivity and photoreflectance of nanocrystalline copper nitride thin films deposited at low temperature. *J. Cryst. Growth* **2005**, *280*, 490–494. [CrossRef]
31. Chen, Y.-H.; Lee, P.-I.; Sakalley, S.; Wen, C.-K.; Cheng, W.-C.; Sun, H.; Chen, S.-C. Enhanced electrical properties of copper nitride films deposited via high power impulse magnetron sputtering. *Nanomaterials* **2022**, *12*, 2814. [CrossRef] [PubMed]
32. Zhang, G.; Lu, Z.; Pu, J.; Wu, G.; Wang, K. Structure and thermal stability of copper nitride thin films. *Indian J. Mater. Sci.* **2013**, *2013*, 725975. [CrossRef]
33. Fan, X.; Wu, Z.; Li, H.; Geng, B.; Li, C.; Yan, P. Morphology and thermal stability of Ti-doped copper nitride films. *J. Phys. D Appl. Phys.* **2007**, *40*, 3430–3435. [CrossRef]
34. Szczęsny, R.; Szłyk, E.; Wiśniewski, M.A.; Hoang, T.K.A.; Gregory, D.H. Facile preparation of copper nitride powders and nanostructured films. *J. Mater. Chem. C* **2016**, *4*, 5031–5037. [CrossRef]
35. Kadzutu-Sithole, R.; Machogo-Phao, L.F.E.; Kolokoto, T.; Zimuwandeyi, M.; Gqoba, S.S.; Mubiayi, K.P.; Moloto, M.J.; Van Wyk, J.; Moloto, N. Elucidating the effect of precursor decomposition time on the structural and optical properties of copper(I) nitride nanocubes. *RSC Adv.* **2020**, *10*, 34231–34246. [CrossRef]
36. Xiao, J.; Li, Y.; Jiang, A. Structure, optical property and thermal stability of copper nitride films prepared by reactive radio frequency magnetron sputtering. *J. Mater. Sci. Technol.* **2011**, *27*, 403–407. [CrossRef]
37. Yuan, X.M.; Yan, P.X.; Liu, J.Z. Preparation and characterization of copper nitride films at various nitrogen contents by reactive radio-frequency magnetron sputtering. *Mater. Lett.* **2006**, *60*, 1809–1812. [CrossRef]
38. Maidul Islam, M.; Georgiev, D.G. Stable stoichiometric copper nitride thin films via reactive sputtering. *Appl. Phys. A* **2022**, *128*, 579. [CrossRef]
39. Reddy, K.V.S.; Reddy, A.S.; Reddy, P.S.; Uthanna, S. Copper nitride films deposited by dc reactive magnetron sputtering. *J. Mater. Sci. Mater. Electron.* **2007**, *18*, 1003–1008. [CrossRef]
40. Scigala, A.; Szłyk, E.; Rerek, T.; Wiśniewski, M.; Skowronski, L.; Trzcinski, M.; Szczesny, R. Copper nitride nanowire arrays—Comparison of synthetic approaches. *Materials* **2021**, *14*, 603. [CrossRef]

MDPI AG
Grosspeteranlage 5
4052 Basel
Switzerland
Tel.: +41 61 683 77 34

Materials Editorial Office
E-mail: materials@mdpi.com
www.mdpi.com/journal/materials

www.ingramcontent.com/pod-product-compliance
Lightning Source LLC
Chambersburg PA
CBHW062052220926
51850CB00010B/2746
* 9 7 8 3 7 2 5 8 7 2 1 6 9 *